Immune Responses and Therapeutics of Inflammatory Bowel Disease

Editor /Author

Salman Assad, M.D.

Shifa Tameer-e-Millat University, Islamabad, Pakistan

Henry Ford Hospital, Michigan, USA

About this Book

This Handbook of on Immune Responses and Therapeutics of IBD provides a complete and comprehensive overview of modern management of IBD in all its complexity, from basic science to gold-standard care.

- Beautifully produced in full color throughout, and with high-quality illustrations, it successfully:
- Provides a solid overview of what clinicians/surgeons/anesthesiologists do, and with topics presented in an order that a clinician will encounter them.
- Presents a holistic look at immunologic responses, foregrounding the interrelationships between team members.

Table of Contents

CHAPTER 1

Innate Immune Response and Gut Microbiota in Inflammatory Bowel Disease

Introduction

Inflammatory bowel disease (IBD), such as Crohn's Disease (CD) and Ulcerative Colitis (UC), are chronic, relapsing inflammatory disorders of the digestive tract resulting from a loss of homeostasis between the intestinal immune system and the gut microbiota in genetically-predisposed individuals [1]. Inappropriate mucosal immune responses, due to dysregulation of tolerance to intestinal microbiota or disruption of the epithelial barrier separating microorganisms from underlying tissues, may contribute to the development or perpetuation of IBD. In fact, genome-wide association studies (GWAS) in patients with IBD have identified more than 150 associated loci, with many of the genetic variants pointing to the importance of barrier function and microbial defense [2], [3], [4]. The new hypothesis for IBD summarizes information obtained from genetic association studies, animal models of inflammation, as well as clinical studies and observations. On the base of these data, IBD are increasingly considered a state of immunodeficiency of the innate arm of immunity, especially for a defective bacterial recognition, autophagy and antigen presentation.

In this context, the intestinal barrier represents a functional unit responsible for two main tasks that are crucial for survival of the individual: allowing nutrient absorption, and defending the body from penetration of unwanted, often dangerous, macromolecules. The gut mucosa is, in fact, a multi-layered system consisting of an external "anatomical" barrier and an inner "functional" immunological barrier. Commensal gut microbiota, the mucous layer, and the intestinal epithelial monolayer constitute the anatomical barrier. The deeper, inner layer consists of a complex network of immune cells organized in a specialized and compartmentalized system known as gut-associated lymphoid tissue or GALT. GALT represents both isolated and aggregated lymphoid follicles and is one of the largest lymphoid organs, containing up to 70% of the body's total number of immunocytes; moreover, it is involved in response to pathogenic microorganisms and provides immune tolerance to commensal bacteria. The ability of GALT to interact with luminal antigens rests on specific mucosal immune cells (*i.e.*, dendritic cells and M-cells), primarily localized to Peyer's patches within the ileum that are intimately positioned at the mucosal-environmental interface and internalize microorganisms and macromolecules. These specialized immune cells have the ability to present antigen to naïve T-lymphocytes, which subsequently produce cytokines and activate mucosal immune responses, when needed. From the intracellular point of view, inflammasomes are a group of protein complexes that assemble upon recognition of a diverse set of noxious stimuli and are now considered the cornerstone of the intracellular surveillance system. Inflammasomes are able to sense both microbial and damage-associated molecular patterns (DAMPs) and initiate a potent innate, anti-microbial immune response [5]. The interaction of these components sustains the maintenance of the delicate equilibrium needed for intestinal homeostasis. Many factors such as alterations in the gut microflora, modifications of the mucus layer and epithelial damage can alter this balance leading to increased intestinal permeability and translocation of luminal contents to the underlying mucosa [6]. The integrity of these structures is necessary for the maintenance of normal intestinal barrier function. Dysregulation of any of the aforementioned components have been implicated not only in the pathogenesis of IBD, but many other GI disorders, including infectious enterocolitis, irritable bowel syndrome, small intestinal bowel overgrowth, and allergic food intolerance [7], [8], [9].

In particular, several lines of evidence have shown that the microbial flora is critical for the development of a normal gut immune system, but can also play a central role in the development of IBD [10], [11], [12], [13]. In support of this concept, the majority of genetically-susceptible murine models of colitis do not develop significant inflammation when raised in a germ-free environment [14], [15], [16], [17], while in others, disease can be attenuated or completely abolished with antibiotic treatment [18], [19]. In this context, innate immune responses that recognize conserved microbial products, such as lipopolysaccharide (LPS) and peptidoglycan [20], are likely to be important in microbial-host interactions and intestinal homeostasis [21] [21]. Critical to the host's sensing of microbes are members of the toll-like receptors (TLR) family that, alone or in combination, recognize a wide array of microbe-associated molecular patterns (MAMPs) on either pathogens or commensals [22]. Furthermore, emerging evidences indicate that intestinal homeostasis and inflammation are driven by cellular elements and soluble mediators that mediate both processes, with several cytokines exhibiting opposing roles depending upon the specific setting. Related to this notion is the dogma that chronic intestinal inflammation characteristic of IBD develops through two distinct phases [23]. Early disease refers to the initial events that take place when homeostatic mechanisms initially fail and acute inflammatory responses cannot be resolved. In contrast, late disease refers to the period when adaptive immunity has been irreversibly primed towards a specific effector phenotype. During these distinct stages of disease progression, innate cytokines play diverse, and often times, dichotomous roles [24], [25].

In the present review, we will comprehensively evaluate the novel and emerging concepts about the role of innate immune response and gut microbiota in controlling mucosal homeostasis and chronic inflammation within the gastrointestinal (GI) tract. On the base of these data, we speculate about the potential implications of the modulation of these factors for treating chronic intestinal inflammation, as well as in designing more efficacious strategies for the prevention and treatment of these devastating GI pathologies.

Gut Microbiota in IBD: cause or consequence?

GI functions are carried out in a dynamic environment inhabited by 1 kg of commensal microbes that include more than 3 mln of genes [26], [27]. They belong to the three domains of life, *Bacteria, Archaea* and *Eukarya* [28], [29], [30], as well as to viral particles [31], [32]. Recent advances in culture-independent molecular techniques, by the analysis of phylogenetic arrays, next generation 16S rRNA sequencing and metagenome sequencing derived from human mucosal biopsies, luminal contents and feces, have shown that four major microbial phyla, (*Firmicutes, Bacteroides, Proteobacteria* and *Actinobacteria*), represent 98% of the intestinal microbiota and fall into three main groups of strict extremophile anaerobes: *Bacteroides, Clostridium* cluster *XIVa* (also known as the Clostridium Coccoides group), and *Clostridium* cluster *IV* (also known as the Clostridium leptum group) [33], [34], [35], [36], [37], [38], [39], [40], [41], [42].

An intricate and mutualistic symbiosis modulates the relationship between the host and the gut microbiota [43], [44], [45]. This relationship is constantly challenged with several factors such as rapid turnover of the intestinal epithelium and overlaying mucus, exposure to peristaltic activity, food molecules, gastric, pancreatic and biliary secretions, defense molecules, drugs, pH and redox potential variations, and exposure to transient bacteria from the oral cavity and esophagus, and can lead to the collapse of the microbial community structure [46]. On the other hand, resident microbes perform several useful functions, including maintaining barrier function, synthesis and metabolism of nutrients, drug and toxin metabolism, and behavioral conditioning [47]. Gut microbiota is also involved in the digestion of energy substrates, production of vitamins and hormones [48], protection from pathogenic bacteria by consuming nutrients and producing molecules that inhibit their growth [49], [50], [51], production of nutrients for mucosal cells [52], [53], [54], augmenting total and pathogen-specific mucosal IgA levels upon infection [55], [56], and in modulating immune system development and immunological tolerance [57].

Unfavorable alteration of microbiota composition, known as dysbiosis, has been implicated in chronic gut, and perhaps also systemic, immune disorders, such as in the pathogenesis of IBD, and other gastrointestinal disorders, including gastritis, peptic ulcer, irritable bowel syndrome [58], [59] and even gastric and colon cancers [60], [61], [62], [63].

Despite the growing evidence that enteric microbes may be involved in the pathogenesis of IBD in genetically susceptible individuals, there has been no convincing evidence reported to suggest that a pathogenic microorganism causes IBD. However, there is a large body of clinical and experimental data implicating the commensal microbiota as a crucial player in the inflammatory responses observed in human and experimental IBD [64]. Of note, the diversion of the fecal stream from a segment of inflamed small bowel is able to reduce intestinal inflammation in patients with CD [65]. On the other side, the restoration of the fecal stream to a segment of surgically resected normal bowel leaded to the induction of intestinal inflammation. This may suggest that components in the fecal stream could induce IBD [66]. Moreover, dysbiosis was found in asymptomatic first-degree relatives of IBD patients, suggesting that dysbiosis may anticipate IBD [67]. Several studies underlined important alterations in the composition of the microbiota in IBD patients [68]. The majority of these studies found a decreased microbial diversity, especially in *Firmicutes* and *Bacteroidetes phyla* [69]. Interestingly, *Faecalibacterium prausnitzii*, a member of the *Firmicutes phyla*, is significantly reduced [70], [71] and possesses a well-documented anti-inflammatory activity [72] [58]. Indeed, *Faecalibacterium prausnitzii* is able to produce butyrate, which has a protective effect on the gut by providing energy to epithelial cells, enhancing mucosal barrier function, increasing intestinal mucous production, stimulating the production of immunosuppressive cytokines, and decreasing the generation of proinflammatory mediators, such as NF-kB [73].

On the other side, Proteobacteria and Actinobacteria are increased in patients with active IBD [74], [75]. Furthermore, specific strains of *Escherichia coli* are augmented in patients with CD [76]. Isolates of these adherent-invasive *E. coli*(AIEC) have been shown to adhere to epithelial cells, and also to invade and replicate within these cells [77]. Also *Clostrium boltae* and *Clostrium symbiosum* are increased in stool samples collected from patients with IBD, but even their role is not clear [78], [79].

Despite these intriguing associations between the microbiota and IBD, however, it is not clear whether alterations in the gut microbiota represent a cause or a consequence of chronic intestinal inflammation. However, several new evidences on the IBD pathogenesis justified the substitution of the immune hyper-reactivity hypothesis with the concept of a defective innate response. In this scenario, gut microbiota acquires a leading role in IBD, which could be a disorder associated with bacterial processing. Indeed, IBD are increasingly considered a state of immunodeficiency of the innate arm of immunity with defective bacterial recognition, autophagy and antigen presentation [80]. Then, the understanding that chronic intestinal inflammation develops through distinct phases and that during these separate stages of disease evolution, similar innate cellular elements and cytokines play diverse, and oftentimes, dichotomous roles has important and clear therapeutic implications.

Gut Microbiota and Innate Immune Response in IBD

In IBD, the pathological relationship between gut microbiota and innate immune response is peculiar and is crucial for triggering and exacerbating the disease. Defective epithelial barrier and increased intestinal permeability can lead to persistent immune activation and have been suggested to play a role in IBD [81]. During normal gut homeostasis, small amounts of luminal antigens translocate across the epithelium both through receptor-mediated endocytosis and non-selective endocytosis. This allows the physiologic sampling of luminal content by the host's immune system [82]. Animal models that lack components of a healthy epithelial barrier have been shown to develop IBD. The lack of N-cadherin in mouse intestinal epithelium has been shown to lead to CD like symptoms [83]. Moreover, alterations in the organic cation transporter (OCTN) gene, which regulates the transport of cationic proteins, such as amino acids and nutrients, are able to produce an augmented susceptibility to CD [84].

Interestingly, IBD patients have antibodies directed against several microbial antigens derived from intestinal bacteria such as *E. coli* and *Pseudomonas fluorescens* as well as yeast (e.g., Saccharomyces). In a scenario of increased gut permeability, invading microorganisms may induce intestinal immune responses that trigger the induction and the exacerbation of IBD [85]. The recent identification of polymorphisms in genes that are involved in intracellular processing and killing of bacteria in patients with CD (NOD2, ATG16L1, and IRGM) suggests, indeed, that inappropriate innate immune responses to luminal bacteria could promote chronic gut inflammation in genetically susceptible individuals [86]. Furthermore, the treatment with certain antibiotics is effective in reducing distal bowel inflammation in IBD patients [87].

The mucous layer that covers the intestinal epithelium represents the first physical barrier that intestinal bacteria and food antigens encounter on the mucosal surface. It provides protection by shielding the epithelium from microorganisms and harmful antigens, while acting as a lubricant for intestinal motility. It consists of two layers: an inner layer and an outer layer. These mucus layers are organized around the highly glycosylated MUC2 mucin, forming a large, net-like polymer that is secreted by the goblet cells. The inner mucus layer is dense and does not allow bacteria to penetrate, thus keeping the epithelial cell surface free from bacteria. The inner mucus layer is converted into the outer layer, which is the habitat of the commensal flora. The outer mucus layer has an expanded volume due to the proteolytic activities provided by the host but probably also caused by commensal bacterial proteases and glycosidases [88]. This compartmentalization seems to be fundamental for the homeostasis in the highly colonized colon. The importance of the mucus barrier was further demonstrated in Muc2 deficient mice where bacteria are in direct contact with the epithelial cells and are also found deep in the crypts as well as inside epithelial cells [89]. Loss of the barrier formed by the inner mucus layer triggers spontaneous colitis and development of colon cancer [90], [91], [92]. Moreover, some CD patients have goblet cell depletion and an impaired mucus layer, which allows bacteria to adhere directly to epithelial cells, and this may contribute to disease progression [93].

The second line of defense against bacteria invasion is formed by the intestinal epithelium, which consists of enterocytes and specialized epithelial cells, such as goblet cells and Paneth cells. Besides the formation of a physical barrier against bacteria, epithelial cells can secrete a number of antimicrobial peptides. Defective expression of antimicrobial peptides has been observed in patients with CD [94]. Paneth cells are a further intestinal epithelial cell [95] type that plays an important role in mucosal homeostasis, and, if functionally impaired, may contribute to IBD [96]. Paneth cells reside at the base of small intestinal crypts and secrete AMPs as well as inflammatory mediators [97].

The epithelium lies between the immune cells in the lamina propria and the microbiota in the gut lumen and it functions to communicate with both. The microbiota signals enterocytes as well as innate cells in the lamina propria via pattern recognition molecules signal receptors, such as cytosolic nucleotide-binding oligomerization domain (NOD)-like receptors (NLRs) and TLRs. These signals have been shown to be necessary for normal homeostasis and resistance to injury [98].

NOD2 is a protein that acts as an intracellular pattern recognition receptor for muramyl dipeptide (MDP), a component of the bacterial wall peptidoglycans. Deficient mice for intracellular pattern recognition receptors, NOD1 and NOD2, had decreased E-cadherin expression with increased epithelial permeability and decreased antimicrobial production [99]. Polymorphisms in the card15/nod2 gene have been constantly associated with increased risk for developing CD [100]. Notably, patients harboring NOD2 risk variants express decreased levels of human α-defensins 5 and 6 (HD5, HD6) in Paneth cells [101]. Although the functional role of NOD2 mutations is still controversial, available evidence suggests that they represent loss-of- function mutations that lead to a reduced activation of nuclear factor kappa-light-chain-enhancer of activated B cells (NF-kB) [102]. Furthermore, Marks et al demonstrated that CD patients may also carry a primary defect in phagocytic function [103].Interestingly, he found that this defect could be independent of the CARD15 polymorphism status, indicating that patients with an intact genotype may still have defective NOD2 signaling. Recent studies in SAMP1/YitFc mice with CD-like ileitis (but intact CARD15 genotype) demonstrate defective MDP-mediated responses, including innate cytokine secretion [104], and give further support to this hypothesis.

Analyzing all the gene mutations associated to IBD, the majority of them fall into few distinct pathophysiological categories, thereby pointing to a limited number of inherent defects [105]. The main part of these abnormalities clearly relate to the function of the innate immune system and effective recognition, intracellular manipulation, and elimination of bacterial factors. These findings strongly support the concept of impaired innate immunity in IBD, thereby leading to defective microbial clearance and persistent antigenic stimulation.

TLRs are expressed on cells within the gut mucosa, including IECs and more broadly, on lamina propria macrophages and dendritic cells [106], [107], [108], [109]. TLR2, -4 and -5 are the major cell-surface sensors of bacterial lipopeptides, including LPS and flagellins, while TLR3, -7, -8 and -9 detect nucleic acid motifs [110]. Among the described TLRs, both human and murine studies have shown the importance of TLR5 and its ligand, the bacterial protein flagellin, which is the major structural component of bacterial flagella, in the regulation of innate and adaptive immune responses that are associated with IBD [111], [112], [113], [114]. However, results from the currently published data do not lead to a definitive conclusion regarding the precise role of TLR5, but open the possibilities to different mechanistic hypotheses. Commensal-derived flagellin has been identified as a dominant antigen in patients with CD, and about 50% of CD patients usually have abnormally high levels of anti-flagellin antibodies that correlate with particular subtypes of severe disease [115] , [116]. Furthermore, a TLR5 stop codon polymorphism, which prevents TLR5 signaling, reduces anti-flagellin adaptive immune responses and appears to be protective against CD in certain ethnic groups [117]. Conversely, some studies have documented a downregulation of TLR5 in human IBD [118] , [119]. In mice, TLR5-deficient animals develop spontaneous colitis in some animal facilities [120] and are more susceptible to dextran sulfate sodium (DSS)-induced colitis, a T-cell independent, chemically-induced model of epithelial damage and acute inflammation, primarily driven by innate immune responses [121]. In contrast, flagellin resulted increased in *T. gondii*-induced ileitis [122]. As such, the role of flagellin/TLR5 in the pathogenesis of IBD is controversial (*i.e.*, pathogenic vs. protective) and the cell types that express TLR5, as well as the TLR5-dependent mechanisms that modulate gut permeability and homeostasis towards gut microbiota remain unclear.

The dichotomous role of Toll/IL-1 Receptor superfamily in IBD

The behavior of the cells mediating innate immunity is altered significantly in individuals withIBD. These cells are orchestrated by specific cytokines, such as those of the interleukin (IL)-1 family. The role of the Toll/IL-1 Receptor (TIR) superfamily and their respective ligands, of which IL-1-like molecules belong, is well established in the pathogenesis of several autoinflammatory and chronic immune disorders [123]. However, the emerging concept that TLRs, as well as IL-1 and its related cytokine family members, also play a critical role in health and maintenance of immune homeostasis is gaining increasing acceptance [124]. The GI system, in fact, represents one of the best examples of where these opposing mechanisms simultaneously take place [125]. A large body of evidence exists that support the contribution of various IL-1 family members, particularly IL-1 and IL-18, to the pathogenesis of IBD, as well as GI-related cancers. However, while selective blockade of proinflammatory cytokines is one of the most effective strategies to down-regulate mucosal inflammation in IBD [126], Phase I clinical trials using strategies to neutralize either IL-1 or IL-18 have failed to show significant efficacy in treating patients with UC and CD, respectively. One potential cause for this failure is the dichotomous functions of these IL-1 family members in inducing disease pathogenesis, while simultaneously promoting protection, within the intestinal gut mucosa.

New insights into the role of cytokine-driven pathways in mucosal immunity have been described, based on several recent studies in animal models of acute intestinal injury, repair, and chronic inflammation. Information derived from these studies reveal that intestinal homeostasis and inflammation are driven by cellular elements and soluble mediators that mediate both processes, with several cytokines exhibiting opposing roles, depending upon the specific setting. This concept is most strongly supported by members of the IL-1 family of cytokines in the pathogenesis of IBD [127] , [128] , [129] , [130] , [131] , [132] , [133] , [134] , [135] , [136] , [137] , [138] , [139] , [140], where the same cytokine can possess both classic pro-inflammatory properties, as well as protective, anti-inflammatory functions, which is primarily dependent on the presence of receptor-bearing cells during the host's disease state.

As such, aside from the established pro-inflammatory properties of IL-1α, IL-1β, IL-18 and their downstream signaling molecules shared with TLR family members, such as NF-kB and myeloid differentiation primary response 88 (MyD88), a growing body of evidence indicates that these mediators are necessary for the maintenance of mucosal homeostasis by effectively handling microbiota, as well as by protecting and restoring the integrity of the epithelial barrier [141] , [142] , [143]. While little is known regarding the potential contributions of other IL-1 family members, such as IL-36, IL-36Ra, IL-37, and IL-38, in chronic intestinal inflammation and gut health, the evolving literature regarding the role of IL-33, the most recently described IL-1 family member is, at present, ambiguous and may reflect yet another example of an innate-type cytokine that possesses multiple functions depending on the immunological status and genetic susceptibility of the host. Although one of the first observations of IL-33-dependent functions in the gut was potent epithelial proliferation and mucus production [144], suggesting the promotion of mucosal repair and healing, dysregulated or uncontrolled IL-33 production may also lead to more pathogenic features characteristic of IBD, including epithelial barrier dysfunction, chronic, relapsing inflammation, and formation of fibrotic lesions [145], [146]. In general, early activation of the intestinal epithelium by pathogenic organisms and/or other noxious environmental antigens elicits the production of epithelial-derived IL-1 family members, including intracellular (ic)IL-1Ra, IL-1α, IL-18, and IL-33. Epithelial disruption often occurs, facilitating translocation of luminal bacterial products and the recruitment of innate immune cells, primarily neutrophils and macrophages that are also a potent source of secreted (s)IL-1Ra, IL-1β, IL-18, and IL-33. Normally, early expression of these mediators dampen acute inflammation and promote epithelial repair and restitution, with the ultimate goal of limiting gut mucosal damage and restoring intestinal homeostasis. Under conditions of either uncontrolled and/or persistent inflammation (*e.g.*, as a result of innate immune dysfunction or host genetic predisposition), infiltration of adaptive immune cells, bearing various IL-1R family members, occurs during the later phases of inflammation, making available an effector population able to respond to IL-1-like ligands. For example, the presence of naïve CD4$^+$ T cells expressing the IL-18R have the ability to respond to IL-18, and in combination with IL-12, represents one of the most potent stimuli for interferon (IFN)γ production and Th1 polarized effector responses, thereby promoting chronic Th1-mediated inflammation. Similar effects can occur upon IL-33 stimulation of naïve CD4$^+$ T cells, but in this case, a robust Th2 immune response results. Furthermore, several levels of regulation exist within each subfamily of IL-1 family members, often including the presence of several agonist isoforms (both precursor and mature, cleaved forms), receptor antagonists, as well as soluble and cell-bound decoy receptors. In addition, the promiscuity of IL-1 family ligands with both binding receptors as well as recruited accessory proteins, instills yet another level of regulation that should be considered when determining the overall biological effects of a specific IL-1 family member agonist. In fact, IL-1 family members cannot be considered in isolation, but with other IL-1-related proteins that can influence their overall interactive effects. An imbalance in the equilibrium between IL-1 family components, dependent on prevalent isoform and receptor binding domain/accessory protein present on effector cells, may be responsible for either driving pathogenic events, including chronic intestinal inflammation, fibrosis, and CRC, or for promoting protection by inducing epithelial repair, mucosal wound healing, and restoration of gut homeostasis (summarized in Figure 1). Based on this new information and the emerging concept that IL-1 family members can possess opposing role in gut health and disease, novel pathogenic hypotheses can be formed with important translational implications in regard to the prevention and treatment of chronic intestinal inflammation, including CD and UC, and CRC.

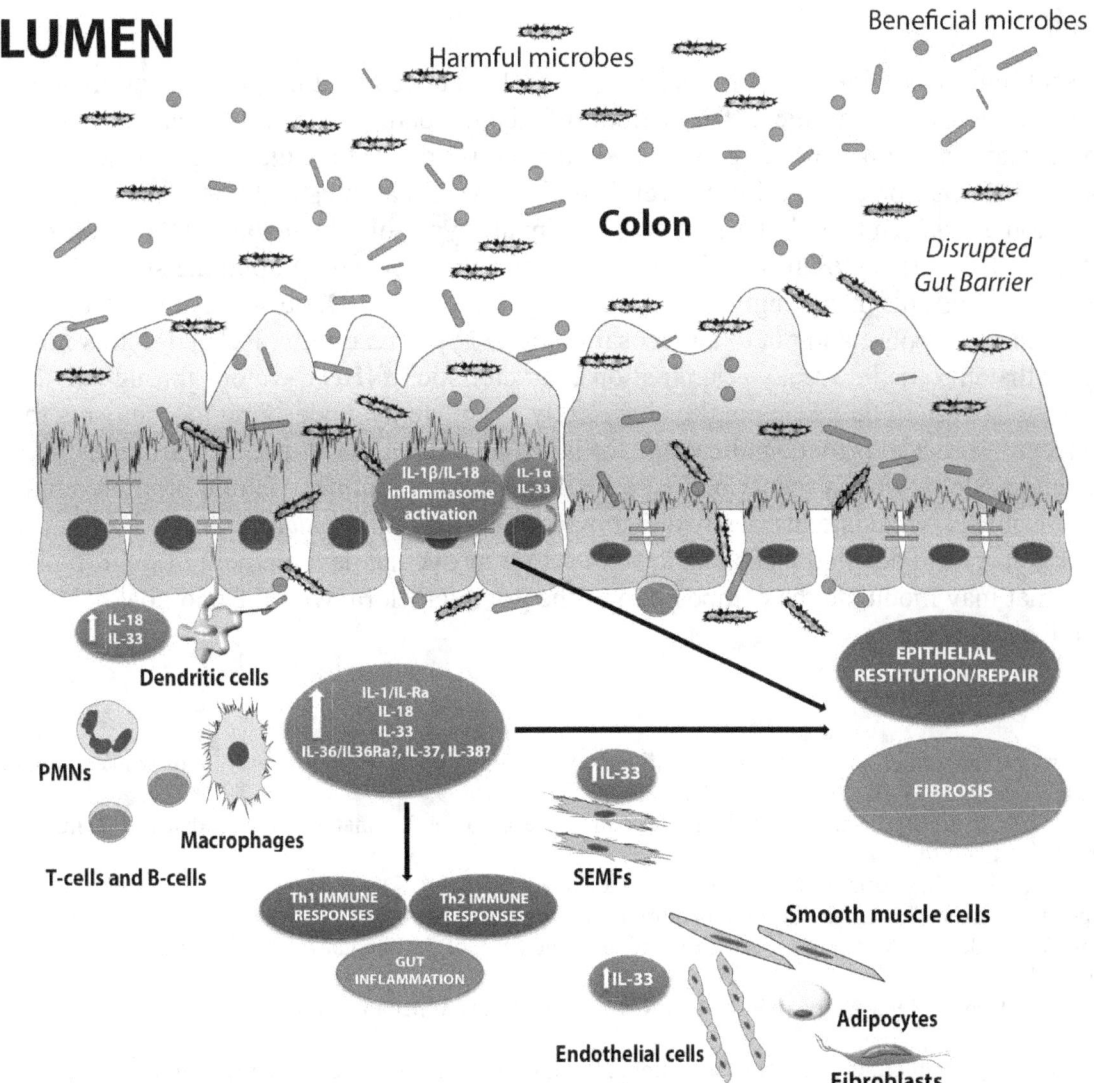

Figure 1. The dichotomous role of IL-1 family members within the gut mucosa. The balance between proinflammatory and protective cytokines is crucial for the maintenance of gut homeostasis. Damage to the epithelium and other proinflammatory stimuli, including PAMPs derived from luminal antigens and the local intestinal microflora, induce the expression of IL-1 family members that are subsequently released by necrotic IECs as potential alarmins (e.g., IL-33 and IL-1α). Depending on the cellular source and presence of receptor-bearing effector cells, IL-1 family members can possess very different and often opposite functions within the gut mucosa. Therapeutic interventions should consider all this processes and whether targeting specific IL-1 family members may be more efficacious during active disease vs. maintaining remission.

Summary

The present review provides an analysis on the interplay between gut microbiota and innate immune response in gut chronic inflammation. In IBD, the pathological relationship between gut microbiota and innate immune response is peculiar and is crucial for triggering and exacerbating the disease. Information derived from several studies reveal that intestinal homeostasis and inflammation are driven by cellular elements and innate-type soluble mediators that mediate both processes, with several cytokines exhibiting opposing roles, depending upon the specific setting and disease phase. This intriguingly concept has suggested that IBD are strictly correlated to defective innate response and not to a mucosal immune hyper-reactivity. Related to this notion is the dogma that chronic intestinal inflammation characteristic of IBD develops through an early and late disease. It will be fascinating to completely elucidate the underlying mechanisms for this new concept in order to provide indications for unexplained patient-to-patient variations in drug efficacy and toxicity. In this scenario, the goal of early phase treatment should promote innate immune response, while late disease will continue to use anti-inflammatory drugs. It will also be important to perform detailed mechanistic studies to improve the development of microbial therapies that may modulate the composition of the gut microflora, with the end goal of promoting gut health.

References

1. Bamias G, Corridoni D, Pizarro TT, Cominelli F. New insights into the dichotomous role of innate cytokines in gut homeostasis and inflammation. Cytokine 2012; 59(3): 451-459.
2. Xavier RJ, Podolsky DK. Unravelling the pathogenesis of inflammatory bowel disease. Nature 2007; 448(7152): 427-434.
3. Jostins L, Ripke S, Weersma RK, Duerr RH, McGovern DP, Hui KY, et al. Host-microbe interactions have shaped the genetic architecture of inflammatory bowel disease. Nature 2012; 491(7422): 119-124.
4. Khor B, Gardet A, Xavier RJ. Genetics and pathogenesis of inflammatory bowel disease. Nature 2011; 474(7351): 307-317.
5. Strowig T, Henao-Mejia J, Elinav E, Flavell R. Inflammasomes in health and disease. Nature 2012; 481(7381): 278-286.
6. Scaldaferri F, Pizzoferrato M, Gerardi V, Lopetuso L, Gasbarrini A. The gut barrier: new acquisitions and therapeutic approaches. J Clin Gastroenterol 2012;4 6 Suppl: S12-17.
7. Camilleri M, Madsen K, Spiller R, Van Meerveld BG, Verne GN. Intestinal barrier function in health and gastrointestinal disease. Neurogastroenterol Motil 2012; 24(6): 503-512.
8. Fasano A. Leaky gut and autoimmune diseases. Clin Rev Allergy Immunol 2012; 42(1): 71-78.
9. Fasano A. Zonulin and its regulation of intestinal barrier function: the biological door to inflammation,autoimmunity, and cancer. Physiol Rev 2011; 91(1): 151-175.
10. Peloquin JM, Nguyen DD. The microbiota and inflammatory bowel disease: Insights from animal models. Anaerobe 2013; 24: 102-106.
11. Scaldaferri F, Petito V, Lopetuso L, Bruno G, Gerardi V, Ianiro G, et al. Pre- and posttherapy assessment of intestinal soluble mediators in IBD: where we stand and future perspectives. Mediators Inflamm 2013; 2013: 391473.
12. Scaldaferri F, Gerardi V, Lopetuso LR, Del Zompo F, Mangiola F, Boskoski I, et al. Gut microbial flora, prebiotics, and probiotics in IBD: their current usage and utility. Biomed Res Int 2013; 2013: 435268.
13. Purchiaroni F, Tortora A, Gabrielli M, Bertucci F, Gigante G, Ianiro G, et al. The role of ntestinal microbiota and the immune system. Eur Rev Med Pharmacol Sci 2013; 17(3): 323-333.
14. Mombaerts P, Mizoguchi E, Grusby MJ, Glimcher LH, Bhan AK, Tonegawa S. Spontaneous development of inflammatory bowel disease in T cell receptor mutant mice. Cell 1993; 75(2): 274-282.
15. Sadlack B, Merz H, Schorle H, Schimpl A, Feller AC, Horak I. Ulcerative colitis-like disease in mice with a disrupted interleukin-2 gene. Cell 1993; 75(2): 253-261.
16. Kuhn R, Lohler J, Rennick D, Rajewsky K, Muller W. Interleukin-10-deficient mice develop chronic enterocolitis. Cell 1993; 75(2): 263-274.

17. Blumberg R, Powrie F. Microbiota, disease, and back to health: a metastable journey. Sci Transl Med 2012; 4(137): 137rv7.

18. Madsen KL, Doyle JS, Tavernini MM, Jewell LD, Rennie RP, Fedorak RN. Antibiotic therapy attenuates colitis in interleukin 10 gene-deficient mice. Gastroenterology 2000; 118(6): 1094-1105.

19. Kang SS, Bloom SM, Norian LA, Geske MJ, Flavell RA, Stappenbeck TS, et al. An antibiotic-responsive mouse model of fulminant ulcerative colitis. PLoS Med 2008; 5(3): e41.

20. Guarino A, Albano F, Ashkenazi S, Gendrel D, Hoekstra JH, Shamir R, et al. European Society for Paediatric Gastroenterology, Hepatology, and Nutrition/European Society for Paediatric Infectious Diseases evidence-based guidelines for the management of acute gastroenteritis in children in Europe: executive summary. J Pediatr Gastroenterol Nutr 2008; 46(5): 619-621.

21. Akira S, Uematsu S, Takeuchi O. Pathogen recognition and innate immunity. Cell 2006; 124(4): 783-801.

22. Bamias G, Corridoni D, Pizarro TT, Cominelli F. New insights into the dichotomous role of innate cytokines in gut homeostasis and inflammation. Cytokine 2012; 59(3): 451-459.

23. Lopetuso LR, Chowdhry S, Pizarro TT. Opposing Functions of Classic and Novel IL-1 Family Members in Gut Health and Disease. Front Immunol 2013; 4: 181.

24. Leser TD, Molbak L. Better living through microbial action: the benefits of the mammalian gastrointestinal microbiota on the host. Environ Microbiol 2009; 11(9): 2194-2206.

25. Neish AS. Microbes in gastrointestinal health and disease. Gastroenterology 2009; 136(1): 65-80.

26. Eckburg PB, Bik EM, Bernstein CN, Purdom E, Dethlefsen L, Sargent M, et al. Diversity of the human intestinal microbial flora. Science 2005; 308(5728): 1635-1638.

27. Gill SR, Pop M, Deboy RT, Eckburg PB, Turnbaugh PJ, Samuel BS, et al. Metagenomic analysis of the human distal gut microbiome. Science 2006; 312(5778): 1355-1359.

28. Scanlan PD, Marchesi JR. Micro-eukaryotic diversity of the human distal gut microbiota: qualitative assessment using culture-dependent and -independent analysis of faeces. ISME J 2008; 2(12): 1183-1193.

29. Zhang T, Breitbart M, Lee WH, Run JQ, Wei CL, Soh SW, et al. RNA viral community in human feces: prevalence of plant pathogenic viruses. PLoS Biol 2006; 4(1): e3.

30. Breitbart M, Haynes M, Kelley S, Angly F, Edwards RA, Felts B, et al. Viral diversity and dynamics in an infant gut. Res Microbiol 2008; 159(5): 367-373.

31. Eckburg PB, Bik EM, Bernstein CN, Purdom E, Dethlefsen L, Sargent M, et al. Diversity of the human intestinal microbial flora. Science 2005; 308(5728): 1635-1638.

32. Gill SR, Pop M, Deboy RT, Eckburg PB, Turnbaugh PJ, Samuel BS, et al. Metagenomic analysis of the human distal gut microbiome. Science 2006; 312(5778): 1355-1359.

33. Hold GL, Pryde SE, Russell VJ, Furrie E, Flint HJ. Assessment of microbial diversity in human colonic samples by 16S rDNA sequence analysis. FEMS Microbiol Ecol 2002; 39(1): 33-39.

34. Backhed F, Ley RE, Sonnenburg JL, Peterson DA, Gordon JI. Host-bacterial mutualism in the human intestine. Science 2005; 307(5717): 1915-1920.

35. Ley RE, Backhed F, Turnbaugh P, Lozupone CA, Knight RD, Gordon JI. Obesity alters gut microbial ecology. Proc Natl Acad Sci U S A 2005; 102(31): 11070-11075.

36. Ley RE, Peterson DA, Gordon JI. Ecological and evolutionary forces shaping microbial diversity in the human intestine. Cell 2006; 124(4): 837-848.

37. Frank DN, St Amand AL, Feldman RA, Boedeker EC, Harpaz N, Pace NR. Molecular-phylogenetic characterization of microbial community imbalances in human inflammatory bowel diseases. Proc Natl Acad Sci U S A 2007; 104(34): 13780-13785.

38. Rajilic-Stojanovic M, Smidt H, de Vos WM. Diversity of the human gastrointestinal tract microbiota revisited. Environ Microbiol 2007; 9(9): 2125-2136.

39. Tap J, Mondot S, Levenez F, Pelletier E, Caron C, Furet JP, et al. Towards the human intestinal microbiota phylogenetic core. Environ Microbiol 2009; 11(10): 2574-2584.

40. Manson JM, Rauch M, Gilmore MS. The commensal microbiology of the gastrointestinal tract. Adv Exp Med Biol 2008; 635: 15-28.

41. Backhed F, Ley RE, Sonnenburg JL, Peterson DA, Gordon JI. Host-bacterial mutualism in the human intestine. Science 2005; 307(5717): 1915-1920.

42. McCracken VJ, Lorenz RG. The gastrointestinal ecosystem: a precarious alliance among epithelium, immunity and microbiota. Cell Microbiol 2001; 3(1): 1-11.

43. Lievin-Le Moal V, Servin AL. The front line of enteric host defense against unwelcome intrusion of harmful microorganisms: mucins, antimicrobial peptides, and microbiota. Clin Microbiol Rev 2006; 19(2): 315-337.

44. Manson JM, Rauch M, Gilmore MS. The commensal microbiology of the gastrointestinal tract. Adv Exp Med Biol 2008; 635: 15-28.
45. Scaldaferri F, Pizzoferrato M, Gerardi V, Lopetuso L, Gasbarrini A. The gut barrier: new acquisitions and therapeutic approaches. J Clin Gastroenterol 2012;4 6 Suppl: S12-17.
46. Sekirov I. RS, Antunes LC, Finlay BB. Gut mircobiota in health and disease. Physiol Rev 2010; 90: 859-904.
47. Silva AM BF, Duarte R, Vieira LQ, Arantes RM, Nicoli JR. Effect of Bifidobacterium longum ingestion on experimental salmonellosis in mice. J Appl Microbiol 2004; 97: 29-37.
48. Truusalu K MRea. Eradication of Salmonella Typhimurium infection in a murine model of typhoid fever with the cimbination of probiotic Lactobacillus fermentum ME-3 and ofloxacin. BMC Microbiol 2008; 8: 132.
49. Searle LE, Best A, Nunez A, Salguero FJ, Johnson L, Weyer U, et al. A mixture containing galactooligosaccharide, produced by the enzymic activity of Bifidobacterium bifidum, reduces Salmonella enterica serovar Typhimurium infection in mice. J Med Microbiol 2009; 58(Pt 1): 37-48.
50. Martens EC RRea. Coordinate regulation of glycan degradation and polysaccharide capsule biosynthesis by a prominent gut symbiont. J Biol Chem 2009; 284: 18445-18457.
51. Burger van Paassen N VAea. The regulation of intestinal mucin MUC2 expression by short chain fatty acid: implications for epithelial pretection. Biochem J 2009; 420: 211-219.
52. Dharmani P, Srivastava V, Kissoon-Singh V, Chadee K. Role of intestinal mucins in annate host defense mechanisms against pathogens. J Innamte Immun 2009; 1: 123-135.
53. Galdeano CM, Perdigon G. The probiotic bacterium Lactobacillus casei induces activation of the gut mucosal immune system through innate immunity. Clin Vaccine Immunol 2006; 13(2): 219-226.
54. Leblanc J, Fliss I, Matar C. Induction of a humoral immune response following an Escherichia coli O157:H7 infection with an immunomodulatory peptidic fraction derived from Lactobacillus helveticus-fermented milk. Clin Diagn Lab Immunol 2004; 11(6): 1171-1181.
55. Allen CA, Torres AG. Host-microbe communication within the GI tract. Adv Exp Med Biol 2008; 635: 93-101.
56. Parashar UD, Gibson CJ, Bresee JS, Glass RI. Rotavirus and severe childhood diarrhea. Emerg Infect Dis 2006; 12(2): 304-306.
57. Scaldaferri F, Nardone O, Lopetuso LR, Petito V, Bibbo S, Laterza L, Gerardi V, Bruno G, Scoleri I, Diroma A, Sgambato A, Gaetani E, Cammarota G, Gasbarrini A. Intestinal gas production and gastrointestinal symptoms: from pathogenesis to clinical implication. Eur Rev Med Pharmacol Sci 2013; 17 Suppl 2: 2-10.
58. Frank DN, St Amand AL, Feldman RA, Boedeker EC, Harpaz N, Pace NR. Molecular-phylogenetic characterization of microbial community imbalances in human inflammatory bowel diseases. Proc Natl Acad Sci U S A 2007; 104(34): 13780-13785.
59. Swidsinski A, Ladhoff A, Pernthaler A, Swidsinski S, Loening-Baucke V, Ortner M, et al. Mucosal flora in inflammatory bowel disease. Gastroenterology 2002; 122(1): 44-54.
60. Hill DA, Artis D. Intestinal bacteria and the regulation of immune cell homeostasis. Annu Rev Immunol 2010; 28: 623-667.
61. Sartor RB. Microbial influences in inflammatory bowel diseases. Gastroenterology 2008; 134(2): 577-594.
62. Rutgeerts P, Goboes K, Peeters M, Hiele M, Penninckx F, Aerts R, et al. Effect of faecal stream diversion on recurrence of Crohn's disease in the neoterminal ileum. Lancet 1991; 338(8770): 771-774.
63. Joossens M, Huys G, Cnockaert M, De Preter V, Verbeke K, Rutgeerts P, et al. Dysbiosis of the faecal microbiota in patients with Crohn's disease and their unaffected relatives. Gut 2011; 60(5): 631-637.
64. Sartor RB. Microbial influences in inflammatory bowel diseases. Gastroenterology 2008; 134(2): 577-594.
65. Chassaing B, Darfeuille-Michaud A. The commensal microbiota and enteropathogens in the pathogenesis of inflammatory bowel diseases. Gastroenterology 2011; 140(6): 1720-1728.
66. Sokol H, Pigneur B, Watterlot L, Lakhdari O, Bermudez-Humaran LG, Gratadoux JJ, et al. Faecalibacterium prausnitzii is an anti-inflammatory commensal bacterium identified by gut microbiota analysis of Crohn disease patients. Proc Natl Acad Sci U S A 2008; 105(43): 16731-16736.
67. Sokol H, Pigneur B, Watterlot L, Lakhdari O, Bermudez-Humaran LG, Gratadoux JJ, et al. Faecalibacterium prausnitzii is an anti-inflammatory commensal bacterium identified by gut microbiota analysis of Crohn disease patients. Proc Natl Acad Sci U S A 2008; 105(43): 16731-16736.
68. Looijer-van Langen MA, Dieleman LA. Prebiotics in chronic intestinal inflammation. Inflamm Bowel Dis 2009; 15(3): 454-462.
69. Sartor RB. Microbial influences in inflammatory bowel diseases. Gastroenterology 2008; 134(2): 577-594.
70. Koboziev I, Reinoso Webb C, Furr KL, Grisham MB. Role of the enteric microbiota in intestinal homeostasis and inflammation. Free Radic Biol Med 2013; 68C: 122-133.

71. Chassaing B, Darfeuille-Michaud A. The commensal microbiota and enteropathogens in the pathogenesis of inflammatory bowel diseases. Gastroenterology 2011; 140(6): 1720-1728.

72. Chassaing B, Rolhion N, de Vallee A, Salim SY, Prorok-Hamon M, Neut C, et al. Crohn disease–associated adherent-invasive E. coli bacteria target mouse and human Peyer's patches via long polar fimbriae. J Clin Invest 2011; 121(3): 966-975.

73. Lozupone C, Faust K, Raes J, Faith JJ, Frank DN, Zaneveld J, et al. Identifying genomic and metabolic features that can underlie early successional and opportunistic lifestyles of human gut symbionts. Genome Res 2012; 22(10): 1974-1984.

74. Lopetuso LR, Scaldaferri F, Petito V, Gasbarrini A. Commensal Clostridia: leading players in the maintenance of gut homeostasis. Gut Pathog 2013; 5(1): 23.

75. Bamias G, Corridoni D, Pizarro TT, Cominelli F. New insights into the dichotomous role of innate cytokines in gut homeostasis and inflammation. Cytokine 2012; 59(3): 451-459.

76. Salim SY, Soderholm JD. Importance of disrupted intestinal barrier in inflammatory bowel diseases. Inflamm Bowel Dis 2011; 17(1): 362-381.

77. Slack E, Hapfelmeier S, Stecher B, Velykoredko Y, Stoel M, Lawson MA, et al. Innate and adaptive immunity cooperate flexibly to maintain host-microbiota mutualism. Science 2009; 325(5940): 617-620.

78. Hermiston ML, Gordon JI. Inflammatory bowel disease and adenomas in mice expressing a dominant negative N-cadherin. Science 1995; 270(5239): 1203-1207.

79. Peltekova VD, Wintle RF, Rubin LA, Amos CI, Huang Q, Gu X, et al. Functional variants of OCTN cation transporter genes are associated with Crohn disease. Nat Genet 2004; 36(5): 471-475.

80. 57

81. Chassaing B, Darfeuille-Michaud A. The commensal microbiota and enteropathogens in the pathogenesis of inflammatory bowel diseases. Gastroenterology 2011; 140(6): 1720-1728.

82. Feller M, Huwiler K, Schoepfer A, Shang A, Furrer H, Egger M. Long-term antibiotic treatment for Crohn's disease: systematic review and meta-analysis of placebo-controlled trials. Clin Infect Dis 2010; 50(4): 473-480.

83. Johansson ME, Larsson JM, Hansson GC. The two mucus layers of colon are organized by the MUC2 mucin, whereas the outer layer is a legislator of host-microbial interactions. Proc Natl Acad Sci U S A 2010; 108 Suppl 1: 4659-4665.

84. Johansson ME, Phillipson M, Petersson J, Velcich A, Holm L, Hansson GC. The inner of the two Muc2 mucin-dependent mucus layers in colon is devoid of bacteria. Proc Natl Acad Sci U S A 2008; 105(39): 15064-15069.

85. Johansson ME, Phillipson M, Petersson J, Velcich A, Holm L, Hansson GC. The inner of the two Muc2 mucin-dependent mucus layers in colon is devoid of bacteria. Proc Natl Acad Sci U S A 2008; 105(39): 15064-15069.

86. Velcich A, Yang W, Heyer J, Fragale A, Nicholas C, Viani S, et al. Colorectal cancer in mice genetically deficient in the mucin Muc2. Science 2002; 295(5560): 1726-1729.

87. Van der Sluis M, De Koning BA, De Bruijn AC, Velcich A, Meijerink JP, Van Goudoever JB, et al. Muc2-deficient mice spontaneously develop colitis, indicating that MUC2 is critical for colonic protection. Gastroenterology 2006; 131(1): 117-129.

88. Larsson JM, Karlsson H, Crespo JG, Johansson ME, Eklund L, Sjovall H, et al. Altered O-glycosylation profile of MUC2 mucin occurs in active ulcerative colitis and is associated with increased inflammation. Inflamm Bowel Dis 2011; 17(11): 2299-2307.

89. Wehkamp J, Harder J, Weichenthal M, Mueller O, Herrlinger KR, Fellermann K, et al. Inducible and constitutive beta-defensins are differentially expressed in Crohn's disease and ulcerative colitis. Inflamm Bowel Dis 2003; 9(4): 215-223.

90. Stanislawowski M, Wierzbicki PM, Golab A, Adrych K, Kartanowicz D, Wypych J, et al. Decreased Toll-like receptor-5 (TLR-5) expression in the mucosa of ulcerative colitis patients. J Physiol Pharmacol 2009; 60 Suppl 4: 71-75.

91. Kaser A, Zeissig S, Blumberg RS. Inflammatory bowel disease. Annu Rev Immunol 2010; 28: 573-621.

92. Bevins CL, Salzman NH. Paneth cells, antimicrobial peptides and maintenance of intestinal homeostasis. Nat Rev Microbiol 2011; 9(5): 356-368.

93. Rakoff-Nahoum S, Paglino J, Eslami-Varzaneh F, Edberg S, Medzhitov R. Recognition of commensal microflora by toll-like receptors is required for intestinal homeostasis. Cell 2004; 118(2): 229-241.

94. Natividad JM, Petit V, Huang X, de Palma G, Jury J, Sanz Y, et al. Commensal and probiotic bacteria influence intestinal barrier function and susceptibility to colitis in Nod1-/-; Nod2-/- mice. Inflamm Bowel Dis 2012; 18(8): 1434-1446.

95. Hugot JP, Chamaillard M, Zouali H, Lesage S, Cezard JP, Belaiche J, et al. Association of NOD2 leucine-rich repeat variants with susceptibility to Crohn's disease. Nature 2001; 411(6837): 599-603.

96. Wehkamp J, Salzman NH, Porter E, Nuding S, Weichenthal M, Petras RE, et al. Reduced Paneth cell alpha-defensins in ileal Crohn's disease. Proc Natl Acad Sci U S A 2005; 102(50): 18129-18134.

97. Wehkamp J, Harder J, Weichenthal M, Schwab M, Schaffeler E, Schlee M, et al. NOD2 (CARD15) mutations in Crohn's disease are associated with diminished mucosal alpha-defensin expression. Gut 2004; 53(11): 1658-1664.

98. Marks DJ, Harbord MW, MacAllister R, Rahman FZ, Young J, Al-Lazikani B, et al. Defective acute inflammation in Crohn's disease: a clinical investigation. Lancet 2006; 367(9511): 668-678.

99. Corridoni D, Kodani T, Rodriguez-Palacios A, Pizarro TT, Xin W, Nickerson KP, et al. Dysregulated NOD2 predisposes SAMP1/YitFc mice to chronic intestinal inflammation. Proc Natl Acad Sci U S A 2013; 110(42): 16999-17004.

100. Bamias G, Corridoni D, Pizarro TT, Cominelli F. New insights into the dichotomous role of innate cytokines in gut homeostasis and inflammation. Cytokine 2012; 59(3): 451-459.

101. Cario E, Podolsky DK. Differential alteration in intestinal epithelial cell expression of toll-like receptor 3 (TLR3) and TLR4 in inflammatory bowel disease. Infect Immun 2000; 68(12): 7010-7017.

102. Gewirtz AT, Navas TA, Lyons S, Godowski PJ, Madara JL. Cutting edge: bacterial flagellin activates basolaterally expressed TLR5 to induce epithelial proinflammatory gene expression. J Immunol 2001; 167(4): 1882-1885.

103. Abreu MT, Arnold ET, Thomas LS, Gonsky R, Zhou Y, Hu B, et al. TLR4 and MD-2 expression is regulated by immune-mediated signals in human intestinal epithelial cells. J Biol Chem 2002; 277(23): 20431-20437.

104. Hershberg RM. The epithelial cell cytoskeleton and intracellular trafficking. V. Polarized compartmentalization of antigen processing and Toll-like receptor signaling in intestinal epithelial cells. Am J Physiol Gastrointest Liver Physiol 2002; 283(4): G833-839.

105. Aderem A, Ulevitch RJ. Toll-like receptors in the induction of the innate immune response. Nature 2000; 406(6797): 782-787.

106. Peloquin JM, Nguyen DD. The microbiota and inflammatory bowel disease: insights from animal models. Anaerobe 2013; 24: 102-106.

107. Lodes MJ, Cong Y, Elson CO, Mohamath R, Landers CJ, Targan SR, et al. Bacterial flagellin is a dominant antigen in Crohn disease. J Clin Invest 2004; 113(9): 1296-1306.

108. Dubinsky MC, Kugathasan S, Mei L, Picornell Y, Nebel J, Wrobel I, et al. Increased immune reactivity predicts aggressive complicating Crohn's disease in children. Clin Gastroenterol Hepatol 2008; 6(10): 1105-1111.

109. Targan SR, Landers CJ, Yang H, Lodes MJ, Cong Y, Papadakis KA, et al. Antibodies to CBir1 flagellin define a unique response that is associated independently with complicated Crohn's disease. Gastroenterology 2005; 128(7): 2020-2028.

110. Dubinsky MC, Kugathasan S, Mei L, Picornell Y, Nebel J, Wrobel I, et al. Increased immune reactivity predicts aggressive complicating Crohn's disease in children. Clin Gastroenterol Hepatol 2008; 6(10): 1105-1111.

111. Targan SR, Landers CJ, Yang H, Lodes MJ, Cong Y, Papadakis KA, et al. Antibodies to CBir1 flagellin define a unique response that is associated independently with complicated Crohn's disease. Gastroenterology 2005; 128(7): 2020-2028.

112. Gewirtz AT, Vijay-Kumar M, Brant SR, Duerr RH, Nicolae DL, Cho JH. Dominant-negative TLR5 polymorphism reduces adaptive immune response to flagellin and negatively associates with Crohn's disease. Am J Physiol Gastrointest Liver Physiol 2006; 290(6): G1157-1163.

113. Stanislawowski M, Wierzbicki PM, Golab A, Adrych K, Kartanowicz D, Wypych J, et al. Decreased Toll-like receptor-5 (TLR-5) expression in the mucosa of ulcerative colitis patients. J Physiol Pharmacol 2009; 60 Suppl 4: 71-75.

114. Ortega-Cava CF, Ishihara S, Rumi MA, Aziz MM, Kazumori H, Yuki T, et al. Epithelial toll-like receptor 5 is constitutively localized in the mouse cecum and exhibits distinctive down-regulation during experimental colitis. Clin Vaccine Immunol 2006; 13(1): 132-138.

115. Vijay-Kumar M, Sanders CJ, Taylor RT, Kumar A, Aitken JD, Sitaraman SV, et al. Deletion of TLR5 results in spontaneous colitis in mice. J Clin Invest 2007; 117(12): 3909-3921.

116. Ivison SM, Himmel ME, Hardenberg G, Wark PA, Kifayet A, Levings MK, et al. TLR5 is not required for flagellin-mediated exacerbation of DSS colitis. Inflamm Bowel Dis 2010; 16(3): 401-409.

117. Erridge C, Duncan SH, Bereswill S, Heimesaat MM. The induction of colitis and ileitis in mice is associated with marked increases in intestinal concentrations of stimulants of TLRs 2, 4, and 5. PLoS One 2010; 5(2): e9125.

118. Dinarello CA. Interleukin-1 in the pathogenesis and treatment of inflammatory diseases. Blood 2011; 117(14): 3720-3732.

119. Lopetuso LR, Chowdhry S, Pizarro TT. Opposing Functions of Classic and Novel IL-1 Family Members in Gut Health and Disease. Front Immunol 2013; 4: 181.

120. Pizarro TT, Cominelli F. Cytokine therapy for Crohn's disease: advances in translational research. Annu Rev Med 2007; 58: 433-444.

121. Rutgeerts P, Vermeire S, Van Assche G. Biological therapies for inflammatory bowel diseases. Gastroenterology 2009; 136(4): 1182-1197.

122. Cominelli F, Nast CC, Clark BD, Schindler R, Lierena R, Eysselein VE, et al. Interleukin 1 (IL-1) gene expression, synthesis, and effect of specific IL-1 receptor blockade in rabbit immune complex colitis. J Clin Invest 1990; 86(3): 972-980.

123. Andus T, Daig R, Vogl D, Aschenbrenner E, Lock G, Hollerbach S, et al. Imbalance of the interleukin 1 system in colonic mucosa–association with intestinal inflammation and interleukin 1 receptor antagonist [corrected] genotype 2. Gut 1997; 41(5): 651-657.

124. Nishiyama T, Mitsuyama K, Toyonaga A, Sasaki E, Tanikawa K. Colonic mucosal interleukin 1 receptor antagonist in inflammatory bowel disease. Digestion 1994; 55(6): 368-373.

125. Ferretti M, Casini-Raggi V, Pizarro TT, Eisenberg SP, Nast CC, Cominelli F. Neutralization of endogenous IL-1 receptor antagonist exacerbates and prolongs inflammation in rabbit immune colitis. J Clin Invest 1994; 94(1): 449-453.

126. Casini-Raggi V, Kam L, Chong YJ, Fiocchi C, Pizarro TT, Cominelli F. Mucosal imbalance of IL-1 and IL-1 receptor antagonist in inflammatory bowel disease. A novel mechanism of chronic intestinal inflammation. J Immunol 1995; 154(5): 2434-2440.

127. Pizarro TT, Michie MH, Bentz M, Woraratanadharm J, Smith MF, Jr., Foley E, et al. IL-18, a novel immunoregulatory cytokine, is up-regulated in Crohn's disease: expression and localization in intestinal mucosal cells. J Immunol 1999; 162(11): 6829-6835.

128. Monteleone G, Trapasso F, Parrello T, Biancone L, Stella A, Iuliano R, et al. Bioactive IL-18 expression is up-regulated in Crohn's disease. J Immunol 1999; 163(1): 143-147.

129. McNamee EN, Masterson JC, Jedlicka P, McManus M, Grenz A, Collins CB, et al. Interleukin 37 expression protects mice from colitis. Proc Natl Acad Sci U S A 2011; 108(40): 16711-16716.

130. Pastorelli L, Garg RR, Hoang SB, Spina L, Mattioli B, Scarpa M, et al. Epithelial-derived IL-33 and its receptor ST2 are dysregulated in ulcerative colitis and in experimental Th1/Th2 driven enteritis. Proc Natl Acad Sci U S A 2010; 107(17): 8017-8022.

131. Beltran CJ, Nunez LE, Diaz-Jimenez D, Farfan N, Candia E, Heine C, et al. Characterization of the novel ST2/IL-33 system in patients with inflammatory bowel disease. Inflamm Bowel Dis 2010; 16(7): 1097-1107.

132. Kobori A, Yagi Y, Imaeda H, Ban H, Bamba S, Tsujikawa T, et al. Interleukin-33 expression is specifically enhanced in inflamed mucosa of ulcerative colitis. J Gastroenterol 2010; 45(10): 999-1007.

133. Seidelin JB, Bjerrum JT, Coskun M, Widjaya B, Vainer B, Nielsen OH. IL-33 is upregulated in colonocytes of ulcerative colitis. Immunol Lett 2010; 128(1): 80-85.

134. Bamias G, Corridoni D, Pizarro TT, Cominelli F. New insights into the dichotomous role of innate cytokines in gut homeostasis and inflammation. Cytokine 2012; 59: 451-459.

135. Sponheim J, Pollheimer J, Olsen T, Balogh J, Hammarstrom C, Loos T, et al. Inflammatory bowel disease-associated interleukin-33 is preferentially expressed in ulceration-associated myofibroblasts. Am J Pathol 2010; 177(6): 2804-2815.

136. Kojouharoff G, Hans W, Obermeier F, Mannel DN, Andus T, Scholmerich J, et al. Neutralization of tumour necrosis factor (TNF) but not of IL-1 reduces inflammation in chronic dextran sulphate sodium-induced colitis in mice. Clin Exp Immunol 1997; 107(2): 353-358.

137. Tebbutt NC, Giraud AS, Inglese M, Jenkins B, Waring P, Clay FJ, et al. Reciprocal regulation of gastrointestinal homeostasis by SHP2 and STAT-mediated trefoil gene activation in gp130 mutant mice. Nat Med 2002; 8(10): 1089-1097.

138. Reuter BK, Pizarro TT. Commentary: the role of the IL-18 system and other members of the IL-1R/TLR superfamily in innate mucosal immunity and the pathogenesis of inflammatory bowel disease: friend or foe? Eur J Immunol 2004; 34(9): 2347-2355.

139. Schmitz J, Owyang A, Oldham E, Song Y, Murphy E, McClanahan TK, et al. IL-33, an interleukin-1-like cytokine that signals via the IL-1 receptor-related protein ST2 and induces T helper type 2-associated cytokines. Immunity 2005; 23(5): 479-490.

140. Lopetuso LR, Scaldaferri F, Pizarro TT. Emerging role of the interleukin (IL)-33/ST2 axis in gut mucosal wound healing and fibrosis. Fibrogenesis Tissue Repair 2012; 5(1): 18.

141. Pastorelli L, De Salvo C, Cominelli MA, Vecchi M, Pizarro TT. Novel cytokine signaling pathways in inflammatory bowel disease: insight into the dichotomous functions of IL-33 during chronic intestinal inflammation. Therap Adv Gastroenterol 2011; 4(5): 311-323.

CHAPTER 2

Short-term Oral Antibiotics Treatment Promotes Inflammatory Activation of Colonic Invariant Natural Killer T and Conventional CD4+ T Cells

Introduction

The gut mucosa is a complex environment, constantly exposed to a vast community of microorganisms, collectively defined as microbiota, which include bacteria, fungi, protozoa, and viruses [1]. Intestinal microbiota establishes a mutualistic relationship with the host, providing metabolic functions and contributing to shape the immune system [2]. Indeed, germ-free mice (GF) manifest profound defects in the immune system development and function [3]. Moreover, the presence of specific bacterial strains in the gut, such as segmented filamentous bacteria and *Clostridia* cluster IV and XIV, have been linked, respectively, to the differentiation and expansion of IL-17 producing CD4[+]Th17 cells and of Foxp3[+] regulatory T cells [4,5]. To preserve intestinal immune homeostasis, active processes must thus operate, aimed at preserving the capacity of the gut-associated immune system to recognize invading pathogens and simultaneously avoiding immune responses against the commensal intestinal microbiota [1].

Intestinal dysbiosis, defined as a perturbation to the structure of intestinal commensal communities, can be triggered by several factors, including a persistent change in dietary habits, gastrointestinal infections, and alcohol misuse [6,7]. Antibiotic administration can also lead to a profound perturbation of intestinal commensal communities, which persists after cessation of therapy, as a consequence of their broad-spectrum of action [8,9]. Clinical evidences support a correlation between antibiotic use and emergence or exacerbation of immune-mediated inflammation, as shown for atopic reactions, asthma, and inflammatory bowel diseases (IBD) [10-12]. Importantly, several studies have consistently demonstrated that antibiotic use early in life predisposes to IBD emergence in Western populations [13].

Despite these evidences, the impact of antibiotic treatment and antibiotic-induced dysbiosis on distinct immune cells functions is still largely unexplored. In this context, few recent data indicate that neonatal mice treated with vancomycin or colistin manifest reduced intestinal lymphoid follicles, while broad-spectrum antibiotics administration diminishes antimicrobial peptides production [14]. Vancomycin treatment additionally decreases Tregs colonic frequencies while prolonged broad-spectrum antibiotics exposure induces intestinal and systemic alterations in the immune repertoire, including memory/effector T cells, Tregs, and dendritic cells [15,16]. At present, however, it is unknown the effect of antibiotic treatment on frequency and functions of intestinal lipid-specific T cells such as iNKT cells.

Invariant natural killer T cells are a subset of CD1d-restricted αβ-T lymphocytes recognizing both self- and microbial-derived glycolipids and showing both innate and adaptive immune characteristics [17-20]. In line with data from conventional CD4[+] T cells, increasing evidences support the existence of mutual mechanisms of regulation between the intestinal microbiota and iNKT cells [21]. During early neonatal and postnatal stages of development, commensal bacteria negatively shape iNKT cell repertoire through a CXCL16-dependent gradient [22]. Additionally, CD1d-dependent lipid antigens isolated from the commensal *B. fragilis* directly influence iNKT cell proliferation and activation status [23].

Here we evaluated the effect of antibiotics and antibiotic-induced microbiota alterations on colonic T cell immune responses, focusing specifically on iNKT cells. We tested the consequences of re-colonization of the gastrointestinal tract with normal or dysbiotic microbiota on iNKT cell phenotype and function and how it might translate into a specific outcome in the absence or presence of intestinal inflammation.

We provide evidences that antibiotic treatment in adult mice profoundly alters frequency and functions of intestinal iNKT cells even in the absence of intestinal inflammation, and that the presence of a dysbiotic microbiota after antibiotic treatment imprints colonic iNKT and conventional CD4+ T cells toward a pro-inflammatory phenotype that altogether contributes to aggravate intestinal inflammation. Nonetheless, the inflammatory potential of the dysbiotic microbiota decreases over time, opening the possibility to temporally intervene to re-equilibrate dysbiosis, thus controlling concomitantly mucosal immune cell activations.

Materials and Methods

Mice

C57BL/6 mice (Charles River, IT) and CXCR6-EGFP/+ mice (purchased as GFP/GFP from JAX, USA, and bred to heterozigosity with C57BL/6 mice) of 8–10 weeks of age were housed at the IEO animal facility in SPF conditions. Animal procedures were approved by the Italian Ministry of Health (Auth. 127/15, 27/13, 913/16).

Experimental Colitis Models

For the induction of acute colitis, mice were given 2% (w/v) DSS (molecular weight 40 kD; TdB Consultancy) in their drinking water for 7 days followed by 2 days of recovery before sacrifice. The weight curve was determined by weighing mice daily. At sacrifice, colons were collected, their length was measured with a caliper, and then divided in portions to be fixed in 10% formalin for histological analyses, snap-frozen for RNA extraction, and for lamina propria mononuclear cells (LPMC) immunophenotyping.

Antibiotic Treatment and Microbiome Reconstitution

To eliminate the gut microflora, mice were administered with a mix of neomycin (1 g/L), ampicillin (1 g/L), vancomycin (0.5 g/L), and metronidazole (1 g/L) in their drinking water. After 14 days of antibiotic treatment, mice received a transfer of mucus (first day) and feces (second and third days) from untreated (eubiotic) or DSS-treated (dysbiotic) donors by oral gavage. Mucus was scraped from donor colons, diluted in PBS, and administered to recipients at 1:1 ratio. Feces were collected from different donor mice, diluted in PBS (50 mg/mL), and administered to recipients by oral gavage. Mice were sacrificed at different time points according to the experimental settings. In some experiments after fecal microbiota transplantation (FMT), acute colitis was induced by administration of 2% DSS.

Cell Isolation

For the isolation of LPMC from colons, Peyer's patches were removed, lamina propria lymphocytes were isolated *via* incubation with 5 mM EDTA at 37°C for 30 min, followed by further digestion with collagenase IV and DNase at 37°C for 1 h. Cells were then separated with a Percoll density gradient (Sigma-Aldrich, St. Louis, MO, USA). Mesenteric LN and spleens were smashed into 70-μm nylon strainers (BD) and eritrocytes lysed with RBC Lysis buffer (BD). Livers were mechanically dissected and mononuclear cells separated with a Percoll density gradient (Sigma-Aldrich, St. Louis, MO, USA).

In some experiments, after isolation, cells were re-stimulated *in vitro* for 3 h with PMA/ionomycin in the presence of Brefeldin A to evaluate cytokine secretion.

Flow Cytometry Analysis

Mouse iNKT cells were identified by CXCR6-[EGFP] expression or by mCD1d:PBS57 Tetramer (NIH Tetramer core facility) staining. Murine cells were stained with combinations of directly conjugated antibodies: CD45.2 (104), CD3 (17A2) CD8α (53–6.7), CD4 (GK1.5), CD69 (H1.2F3), CD19 (1D3), CD11b (M1/70), F4/80 (BM8), Ly6g (1A8), Ly6c (AL-21), CD11c (HL3) all purchased from BD, eBioscience, or Biolegend. Gating strategy to identify T cells comprised the exclusion of CD11b$^+$, CD19$^+$, and CD11c$^+$ cells (defined as "lineage").

Intracellular staining of cytokines was performed according to standard methods. Cells were fixed and permeabilized with Cytofix/Cytoperm (BD) before addition of the following antibodies: anti-IFNg (XMG1.2), anti-IL17A (TC11-18H10.1), anti-IL10 (JES5-16E3), anti IL22 (Poly5164) (BD or eBioscience or Biolegend). Samples were analyzed by a FACSCanto flow cytometer (BD), gated to exclude nonviable cells. Data were analyzed using FlowJo software (Tristar).

qPCR Protocol for Quantification of Tissue mRNA

Total RNA from mouse colonic tissues was isolated using TRIZOL and Quick-RNA MiniPrep (ZymoResearch) according to manufacturer's instructions. cDNAs were generated from 1 μg of total RNA with reverse transcription kit (Promega). Gene expression levels were evaluated by qPCR and normalized to *Rpl32* gene expression. The primer sequences are collected in Table S1 in Supplementary Material.

16S qPCR Protocol for Quantification of Bacterial DNA

Total bacterial DNA was extracted from mouse feces with the G'NOME DNA extraction kit (MP Biomedicals) according to manufacturer's specifications. Bacterial DNA was analyzed by qPCR using 16S rDNA primers (SIGMA) and collected in the Table S1 in Supplementary Material. The relative abundance of each bacterial group was normalized to Eubacteria using the $2^{-\Delta\Delta Ct}$ method.

Histological Analysis

Tissue processing was performed with a LEICA PELORIS processor before paraffin embedding. Murine samples were included using an automated system (SAKURA Tissue-Tek). After Hematoxylin and Eosin staining, snapshots of histology were taken using an Aperio CS2 microscope with a scanning resolution of 50,000 pixels per inch (0.5 μm per pixel with 10×

objective and 2.5 μm per pixel when scanning at 4×). Scoring of disease activity was performed according to the criteria described in Table S2 in Supplementary Material.

Statistical Analysis

Statistical significance was calculated using Kruskal–Wallis nonparametric test for multiple comparisons or unpaired nonparametric Mann–Whitney t-test for comparisons between two groups. $P < 0.05$ (*), $P < 0.01$ (**) $P < 0.001$ (***) were regarded as statistically significant. Outliers detected with Grubb's test.

Results

Antibiotic Treatment Influences Colonic iNKT Cell Frequency and Phenotype

To assess the effects of antibiotic treatment on intestinal iNKT cells phenotype and function under homeostatic conditions, CXCR6-[EGFP/+] mice were treated for 2 weeks with a broad-spectrum antibiotic cocktail (ABX, vancomycin, metronidazole, ampicillin, penicillin), targeting both aerobic and anaerobic bacteria (Figure 1A). CXCR6-[EGFP/+] mice are a useful strain to track iNKT cells [24]. The complete depletion of intestinal microbiota by ABX treatment was confirmed by colony forming unit analysis and qPCR analyses (data not shown). Subsequently, mice were either maintained in ABX treatment (ABX, upper scheme) or reconstituted by oral gavage with mucosa-associated and fecal bacteria (FMT; Figure 1A, middle scheme). To evaluate whether the characteristics of the microbiota utilized to reconstitute the gastrointestinal tract upon ABX treatment might have a subsequent impact on iNKT cells frequency and function, mucus and feces derived from healthy mice (eubiotic FMT) or from mice with intestinal dysbiosis (dysbiotic FMT) were engrafted to ABX-treated mice (Figure 1A, middle scheme). Dysbiotic microbiota was obtained from mice with acute DSS colitis, as we and others have observed that this condition is associated with a relevant alteration of both fecal and mucosa-associated microbial composition (Figure 1B)[25]. Histological evaluation of colonic tissues (Figure 1C) and qPCR analyses of tissue derived mRNA (Figure 1D) revealed that neither antibiotic treatment nor recolonization with eubiotic or dysbiotic microorganisms induced macroscopic changes in the colonic tissue architecture (Figure 1C) and all treatments failed to upregulate inflammatory genes (Figure 1D).

FIGURE 1

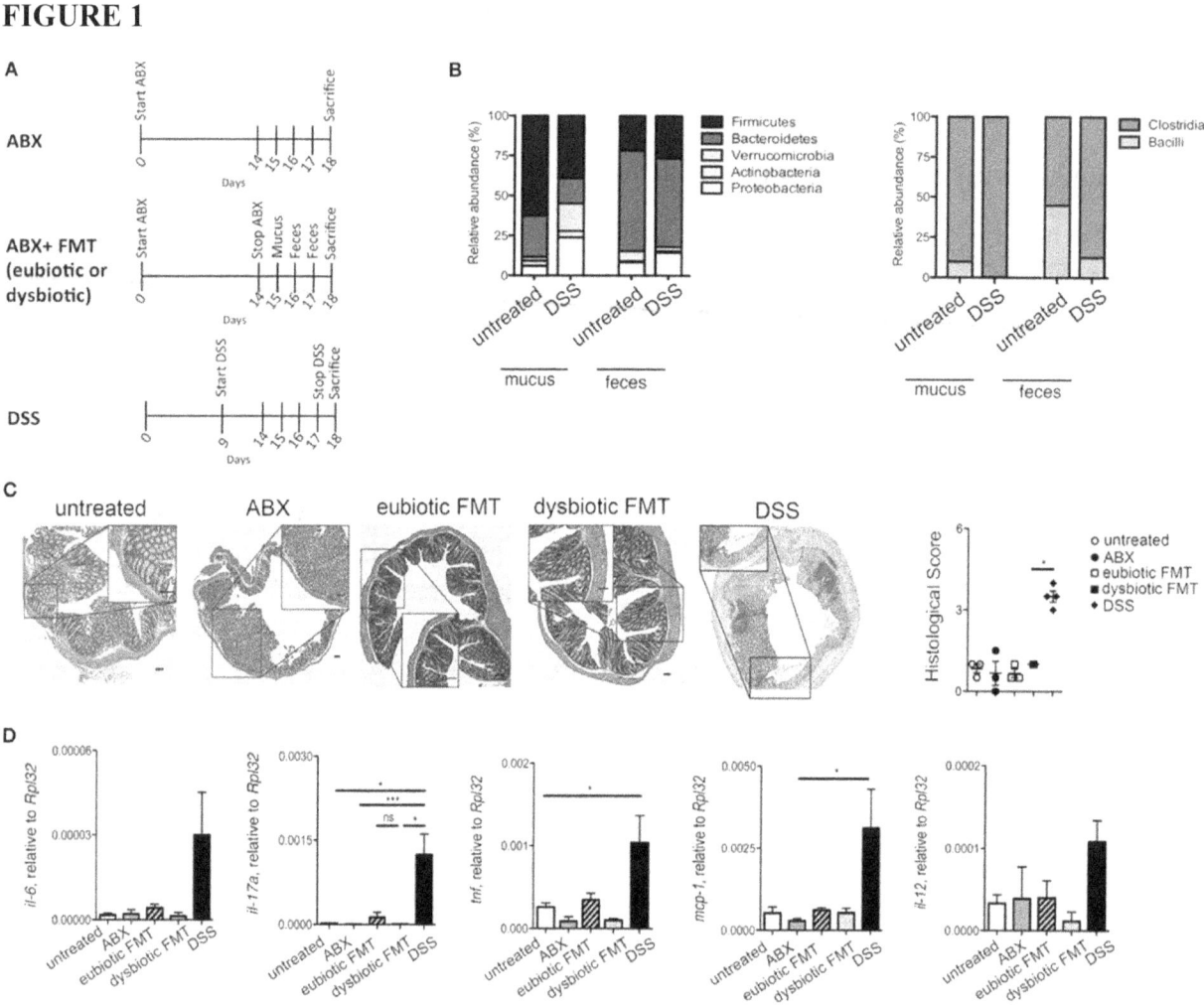

Figure 1. Antibiotic treatment does not alter mucosal architecture under homeostatic conditions. **(A)** Schematic representation of the treatments. **(B)** Abundances of dominant mucus-associated and fecal bacterial groups (phyla level on the left and class level among *Firmicutes* on the right) derived from untreated and DSS-treated mice and utilized to recolonize ABX-treated mice **(C)** H&E staining and cumulative histological score on colon specimens of untreated (open circles), antibiotic treated (ABX, closed circles), transplanted with eubiotic (open squares) or dysbiotic fecal microbiota transplantation (FMT) (closed squares), and of DSS-treated mice (closed diamonds). Scale bar 100 μm. **(D)** Colonic expression levels of *Il6, il17a, tnf, mcp-1, il-12* in untreated (white bars), ABX-treated(gray bars), reconstituted with eubiotic (striped bars) or dysbiotic (dotted bars) FMT or in DSS-treated mice (black bars). Significance was determined using Kruskal–Wallis nonparametric test and expressed as mean SEM untreated *n* = 8, ABX-

treated $n = 10$, reconstituted with eubiotic FMT $n = 9$, with dysbiotic FMT $n = 11$, or in DSS-treated $n = 11$ mice in four independent experiments. Outliers detected with Grubb's test. $P < 0.05$ (*), $P < 0.001$ (***) were regarded as statistically significant.

In sharp contrast, treatment of adult mice with broad spectrum antibiotics was sufficient to induce a significant expansion of iNKT cells in the colon (Figures 2A–C; Figures S1 and S2 in Supplementary Material). Importantly, reconstitution of the gut microbiota with a eubiotic FMT restored iNKT cell frequency (Figures 2A–C), a phenomenon that was not observed upon microbial reconstitution with the microbiota derived from DSS-treated mice. In contrast with these data, CD4$^+$ T cells accumulation in the colon was unaffected by antibiotic treatment or by microbiota recolonization, regardless its origin (Figures 2A–C).

FIGURE 2

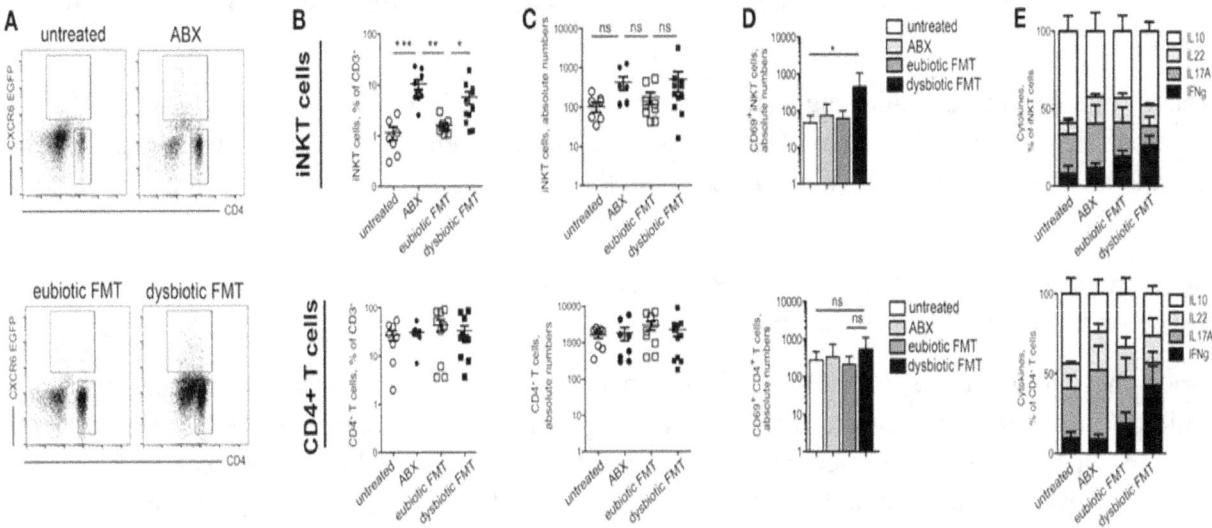

Figure 2. Antibiotic treatment influences colonic invariant natural killer T (iNKT) cell frequency and function **(A–C)** representative dot plots **(A)**, cumulative frequency **(B)**, and absolute numbers **(C)** of iNKT cells (upper panels) and CD4$^+$ T cells (lower panels) in untreated mice (open circles), ABX-treated mice (closed circles), mice reconstituted with eubiotic fecal microbiota transplantation (FMT) (open squares), or with microbiota from DSS-treated mice (dysbiotic FMT, closed squares). **(D)** Absolute numbers of CD69$^+$ cells among iNKT cells (upper panels) and CD4$^+$ T cells (lower panels) in untreated mice (white bars), ABX-treated mice (light gray bars), mice reconstituted with eubiotic FMT (dark gray bars), or with microbiota from DSS-treated mice (dysbiotic FMT, black bars) **(E)** Cytokine production by iNKT cells (upper panels) and CD4$^+$ T cells (lower panels) in untreated, ABX-treated, reconstituted with eubiotic or with dysbiotic FMT. Histograms normalized to 100% of production of total cytokines. Significance was determined using Kruskal–Wallis nonparametric test and expressed as mean SEM. Untreated $n = 8$, ABX-treated $n = 10$, reconstituted with eubiotic FMT $n = 9$, with dysbiotic FMT $n = 11$ or in DSS-treated $n = 11$ mice in four independent experiments. Outliers

detected with Grubb's test. $P < 0.05$ (*), $P < 0.01$ (**), $P < 0.001$ (***) were regarded as statistically significant.

The expression of CD69 (Figure 2D), a surface molecule associated to T cells functional activation, and the cytokine profile (Figure 2E) of both iNKT cells and CD4$^+$ T cells isolated from ABX-treated mice did not significantly differ from that of cells isolated from mice reconstituted with an eubiotic microflora. Conversely, re-colonization of ABX-treated mice with a dysbiotic microbiota was sufficient to upregulate CD69 and to stimulate IFNg secretion by both iNKT cells (Figures 2D,E), and by CD4$^+$ colonic T cells.

Interestingly, these effects were observed in iNKT cells isolated from the colon, but not from the mesenteric LNs (mLN) or from the spleens (Figure S3 in Supplementary Material) of treated mice, thus confirming that the tight regulation of iNKT cells frequency and functions operated by the commensal microbiota occurs in the colonic microenvironment rather than in the periphery [23].

Taken together, these results suggest that expansion of colonic iNKT cells and CD4$^+$ T cells are differentially modulated upon antibiotic treatment in the absence of intestinal inflammation, and that the microbiota composition influences both the accumulation and the functional activation of colonic iNKT cells.

Recolonization of Antibiotic-Treated Mice with a Dysbiotic Microbiota Aggravates Subsequent Intestinal Inflammation

Since the nature of the microbiota utilized to reconstitute the gastrointestinal tract following antibiotic treatment exerted an influence over expansion and cytokine profile of colonic iNKT cells, we next assessed whether these effects translated into a parallel effect over experimental intestinal inflammation. To address this issue, mice were treated with broad-spectrum ABX for 2 weeks before reconstitution with either eubiotic or dysbiotic microbiota. Shortly after re-colonization, acute colitis was induced by administration of DSS in their drinking water (Figure 3A). Of note, mice recolonized with a dysbiotic microbiota exhibited a more severe colitis (as indicated by a more profound weight loss, Figure 3B, and a higher histological score, Figure 3D) than those reconstituted with a eubiotic microbiota. No difference was found in colon length (Figure 3C). Reconstitution with a dysbiotic microflora was sufficient to upregulate colonic *ifng* (Figure 3E).

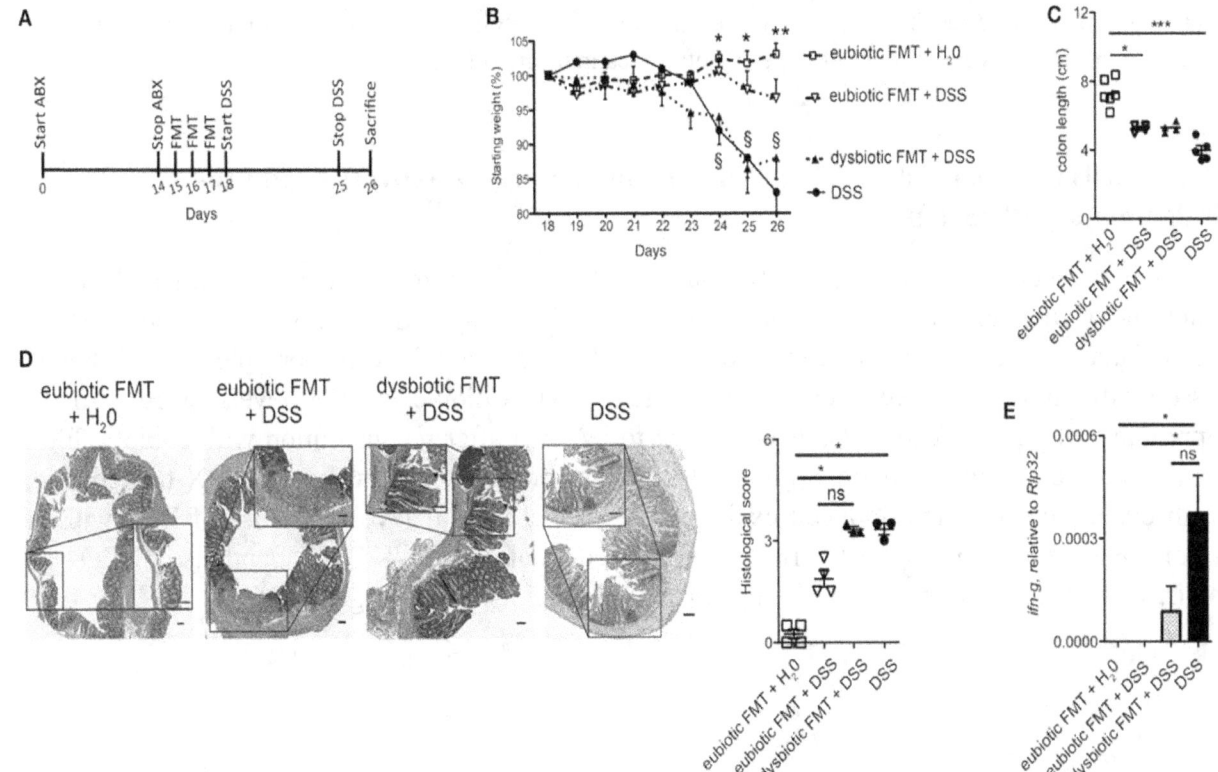

FIGURE 3

Reconstitution of ABX-treated mice with dysbiotic microbiota aggravates subsequent intestinal inflammation. (A) Schematic representation of the treatment. (B) Weight loss, (C) colon length, and (D) histological evaluation of mice treated with eubiotic FMT before water administration or treated with normal or dysbiotic fecal microbiota transplantation (FMT) before DSS administration, or treated with DSS without prior exposure to antibiotics. *Statistical analysis between normal FMT + H20 and DSS; §Statistical analysis between eubiotic FMT + H20 and dysbiotic FMT + DSS performed with Mann–Whitney test. (D) Shows H&E staining on colon specimens. Scale bar 100 μm. (E) Colonic expression levels of ifng in mice reconstituted with eubiotic microbiota and unchallenged (gray bars), reconstituted with eubiotic bacteria and DSS-treated (striped bars), reconstituted with dysbiotic microbiota and DSS-treated (dotted bars) and DSS-treated (black bars). Eubiotic FMT + H20 n = 6, eubiotic FMT + DSS n = 4, dysbiotic FMT + DSS n = 4, or DSS-treated n = 5 mice, two independent experiments. Significance was determined using Kruskal–Wallis nonparametric test and expressed as mean SEM Outliers detected with Grubb's test. P < 0.05 (* and §), P < 0.01 (** and §), P < 0.001 (***) were regarded as statistically significant.

Taken together, these results suggest that presence of an underlying dysbiotic microbiota may exert a negative influence over subsequent experimental intestinal inflammation.

iNKT Cells Exposed to a Dysbiotic Microbiota Acquire an Activated and Pro-inflammatory Phenotype

We next evaluated if the observed negative effects of the re-colonization with a dysbiotic microbiota after antibiotic treatment were associated to a specific phenotype of colonic iNKT cells. First, the colonic expression levels of cxcl16, the chemokine responsible for iNKT cells tissue attraction (22), were evaluated (Figure 4A). Colonic cxcl16 levels were strongly upregulated by DSS treatments, as compared to its level after reconstitution with eubiotic FMT in the absence of inflammation (Figure 4A). Noteworthy, reconstitution of ABX-treated mice with dysbiotic microbiota-induced cxcl16 expression at similar levels as those of DSS without prior ABX treatement. Consistently, a marked accumulation of iNKT cells, and also of CD4+ T cells, was observed in the colon of mice upon DSS administration (Figure 4B).

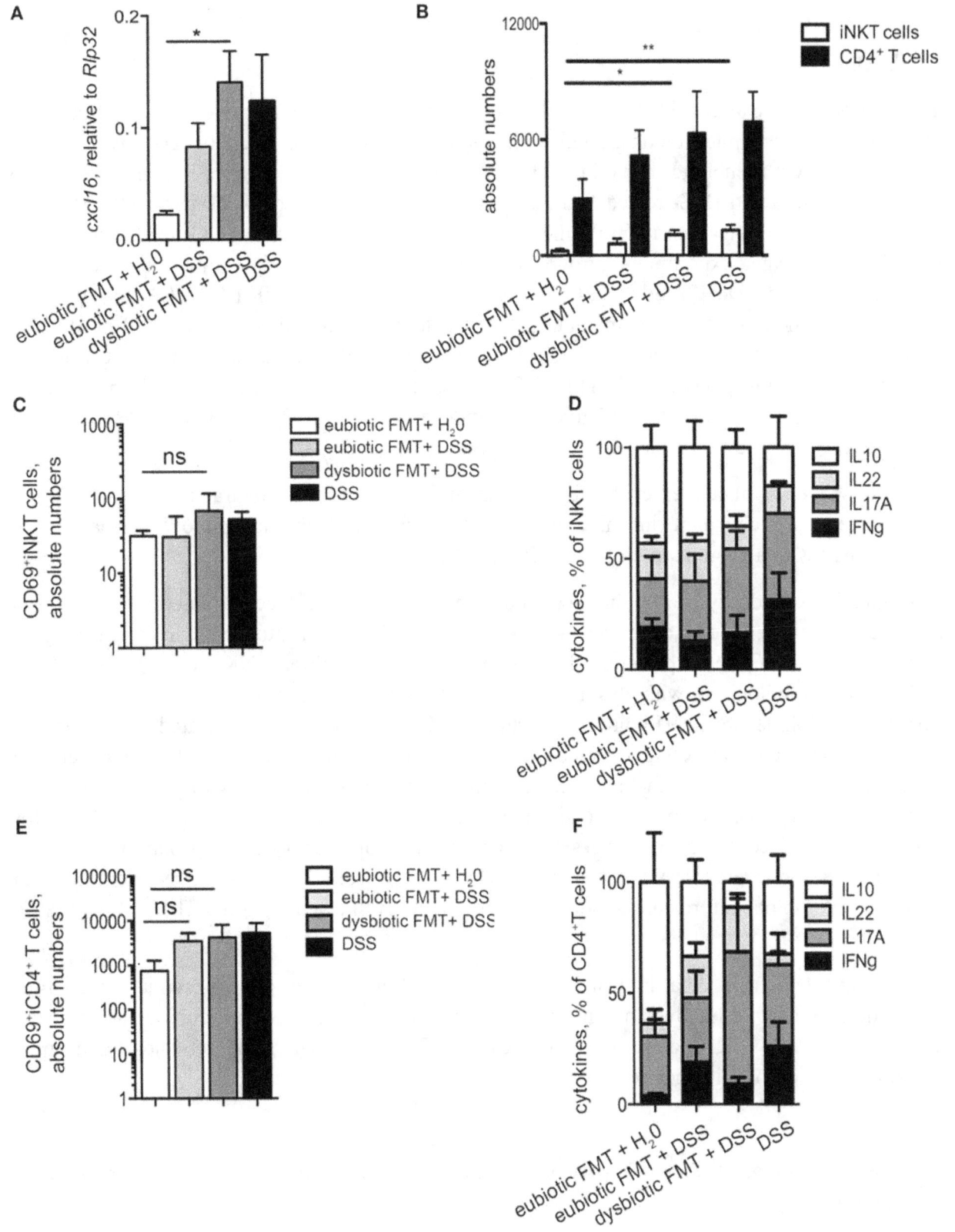

FIGURE 4

T cell responses are affected by the origin of the transplanted microbiota. (A) Colonic expression levels of cxcl16 in mice reconstituted with eubiotic bacteria and unchallenged (white bars), reconstituted with eubiotic bacteria and DSS-treated (light gray bars), reconstituted with dysbiotic bacteria and DSS-treated (dark gray bars) and DSS-treated (black bars). (B) Absolute numbers of colonic invariant natural killer T (iNKT) cells (white bars) and CD4+ T cells (black bars) in the indicated experimental groups. (C,E) CD69 absolute numbers of colonic iNKT cells (C) and of colonic CD4+ T cells (E). (D,F) Cytokine production by iNKT cells (D) and of CD4+ T cells (F) in the indicated experimental groups. Histograms normalized to 100% of production of total cytokines. Significance was determined by Kruskal–Wallis nonparametric test. Eubiotic fecal microbiota transplantation (FMT) + H20 n = 6, eubiotic FMT + DSS n = 4, dysbiotic FMT + DSS n = 4, or DSS-treated n = 5 mice, two independent experiments. Outliers detected with Grubb's test. $P < 0.05$ (*), $P < 0.01$ (**) were regarded as statistically significant.

As a consequence of higher cxcl16 expression and increased chemoattraction, colonic iNKT cells were more abundant in mice reconstituted with a dysbiotic microbiota than those reconstituted with a eubiotic microbiota (Figure 4B).

Additionally, a tendency of a higher accumulation of colonic iNKT cells expressing CD69 was observed in colitic mice re-colonized with a dysbiotic microbiota after ABX treatment (Figure 4C). Similarly to CD69 expression, the cytokine profile of colonic iNKT cells isolated from colitic mice re-colonized with dysbiotic microbiota was skewed toward a pro-inflammatory Th1/Th17 cytokine profile (Figure 4D), which in CD4+ T cells is associated to pathogenic properties (26). On the contrary, re-colonization of the gastrointestinal tract with a eubiotic microbiota did not sustain the activated/inflammatory phenotype of iNKT cells, but rather maintained them toward an uninflamed/IL10-secreting cytokine profile (Figure 4D). As for colonic CD4+ T cells, no differences were observed among DSS-treated groups in terms of CD69 expression (Figure 4E), although also CD4+ T cells isolated from mice re-colonized with dysbiotic microbiota before colitis induction were similarly skewed toward a Th1/Th17 cytokine profile (Figure 4F).

Taken together, these data indicate that re-colonization of the gastrointestinal tract after antibiotic treatment with a dysbiotic microbiota, but not with an eubiotic microbiota, sustains the accumulation of iNKT cells (and of CD4+ T cells) with an activated and pro-inflammatory/colitogenic phenotype.

Dysbiotic Microbiota Effects after Antibiotic Treatment on Colonic iNKT Cells Are Time-Dependent

We next aimed to evaluate whether the negative effects of re-colonization with dysbiotic microbiota over intestinal inflammation were persistent over time (Figure 5A). To this end, mice were treated with ABX for 2 weeks and then reconstituted with either a eubiotic or a dysbiotic microbiota. This time, DSS administration was performed 1 week after the last microbiota gavage (Figure 5A). By following this protocol, no differences could be observed between the effects of eubiotic and dysbiotic re-colonization after antibiotic treatment in terms of weight loss (Figure 5B), colon length reduction (Figure 5C), histological score (Figure 5D), and expression of inflammatory genes in the colonic tissue (Figure 5E).

FIGURE 5

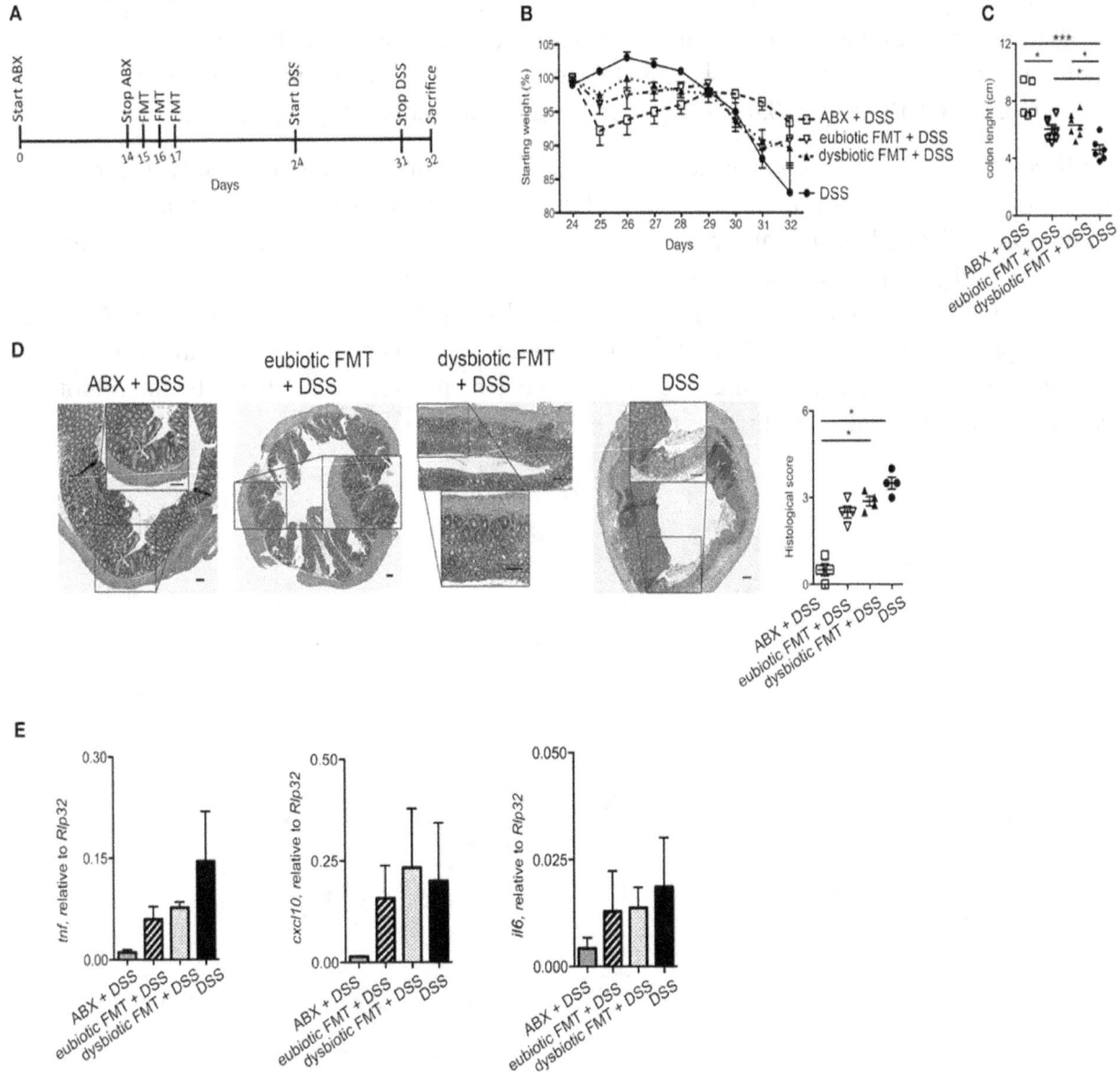

FIGURE 5

Dysbiotic fecal microbiota transplantation (FMT) effects are time-dependent (A) Schematic representation of treatment. (B) Weight loss, (C) colon length, and (D) histological evaluation of mice treated with ABX before DSS administration or treated with eubiotic or dysbiotic FMT before DSS administration or with DSS. (D) Shows H&E staining on colon specimens. Scale bar 100 μm. (E) Colonic expression levels of tnf, cxl10, and il6 in ABX-treated mice before DSS treatment (gray bars), reconstituted with eubiotic bacteria and DSS-treated (striped bars), reconstituted with dysbiotic bacteria and DSS-treated (dotted bars) and DSS-treated (black bars). Significance determined using Kruskal–Wallis nonparametric test and expressed as mean SEM P < 0.05 (*), P < 0.01 (**), P < 0.001 (***) were regarded as statistically significant.

Similarly, the difference in colonic cxcl16 expression between mice re-colonized with eubiotic or dysbiotic microbiota was abolished when a longer time was allowed before colitis induction (Figure 6A), mirrored by similar recruitment and abundance of colonic iNKT cells among the two groups (Figure 6B). A similar re-equilibration between the effects of eubiotic and dysbiotic reconstitution before colitis induction was observed also on the activation status (Figure 6C) and cytokine profile (Figure 6D) of iNKT cells. To note, the differences of pro-inflammatory phenotype induction on conventional CD4+ T cells between eubiotic and dysbiotic microbiota exposure were also abolished (Figures 6E,F).

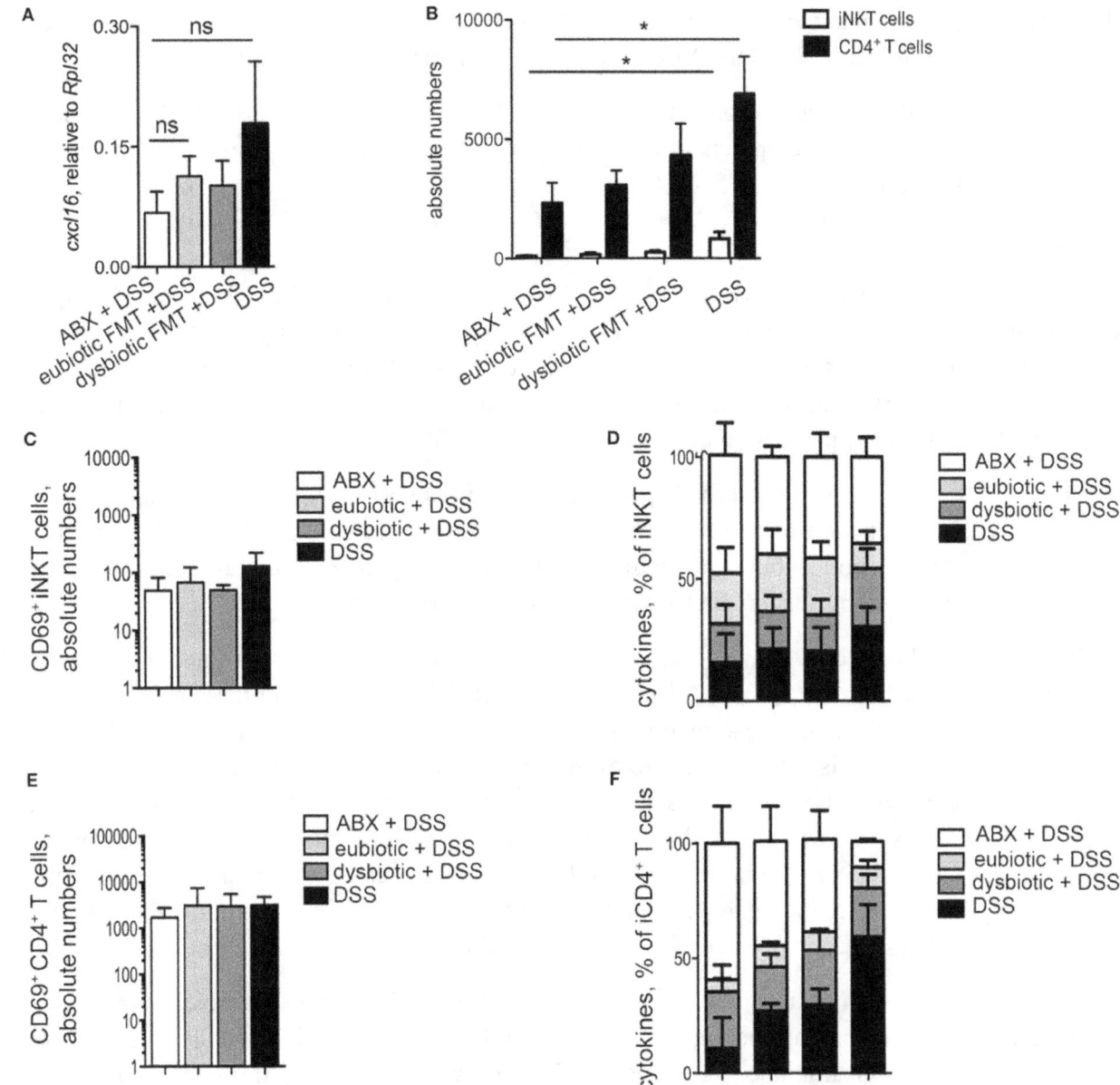

FIGURE 6

Dysbiotic microbiota effects after antibiotic treatment on colonic invariant natural killer T (iNKT) cells are time-dependent. (A) Colonic expression levels of cxcl16 in mice treated with antibiotics and treated with DSS (white bars), transplanted with eubiotic bacteria and DSS-treated (light gray bars), transplanted with dysbiotic bacteria, and DSS-treated (dark gray bars) and DSS-treated (black bars). (B) Absolute numbers of colonic iNKT cells (white bars) and CD4+ T cells (black bars) in the indicated experimental groups. (C,E) CD69 absolute numbers of colonic iNKT cells (C) and of colonic CD4+ T cells (E). (D,F) Cytokine production by iNKT cells (D) and CD4+ T cells (F) in the indicated experimental groups. Histograms normalized to 100% of production of total cytokines. Significance determined using Kruskal–Wallis nonparametric test and expressed as mean SEM. ABX + DSS n = 5, eubiotic FMT + DSS n = 6, dysbiotic FMT + DSS n = 7, DSS-treated n = 6 mice, two independent experiments. Outliers detected with Grubb's test. P < 0.05 (*) were regarded as statistically significant.

Taken together, these results confirm that the pro-inflammatory effects of the dysbiotic microbiota over colitis development might be significantly affected by the time allowed for microbial reconstitution.

Discussion

The immune system and the host microbiota shape each other throughout life, generating an equilibrium that is daily challenged as a result of exposure to pathogens, to environmental factors, or to dietary changes [1]. Recently, antibiotic usage has been identified as one important trigger of disequilibrium between the immune system and the intestinal microbiota [8], with still poorly understood consequences on the overall human health.

We here report that a short-term broad-spectrum antibiotic administration is sufficient to affect the phenotype and function of colonic iNKT cells and that the re-colonization of the intestinal microbiota with dysbiotic, but not with eubiotic, microorganisms aggravates subsequent intestinal inflammation by sustaining T cells pro-inflammatory phenotype.

A healthy microbial ecosystem is defined by its richness and its resistance to external perturbations. Data from the Human Microbiome Project [27] indicate that multiple eubiotic states of the microbial ecosystem, whose taxonomic compositions are influenced by geography, age, and dietary habits, co-exist throughout life even without manifest signs of disease [7]. Dysbiosis can be defined as any short-term or stable alteration in the microbial ecology affecting the taxonomical composition as well as the metagenomic functions of the microbial community [7]. Growing evidences support the notion that antibiotic exposure cause short-term and long-term intestinal dysbiosis by reducing or completely removing normally residing members of the microbiota, as a consequence of microbial killing or diminished bacterial proliferation [8, 28, 29].

Even though it has been extensively shown that microbiota absence in GF mice impacts the immune system development and function [3, 30], the effects of antibiotic-dependent microbiota ablation on the mucosal immune system has so far been poorly studied. Antibiotic exposure in early life reduces bacterial diversity and alters the composition of the commensal microbiota (8), a phenomenon linked to increased susceptibility to develop immune-related pathologies in adult life such as allergic inflammation and asthma [10, 11].

A recent report indicate that prolonged (8 weeks) broad spectrum ABX usage [16] induces a reduction in the relative abundance of several immune cell populations, including T cells, both in periphery and in the intestinal tissues. This event can be reverted by re-colonization of antibiotic-treated mice with a normal microflora.

Here, we show that also a shorter exposure to a broad-spectrum ABX cocktail, a schedule of treatment closer to the ones used in clinical practice is sufficient to induce alterations in the relative abundances and cytokine profile of lipid-specific iNKT cells and of conventional CD4+ T cells. Interestingly, 2 weeks ABX administration does not induce macroscopic signs of inflammation, as demonstrated by unaltered colonic architecture and low expression of inflammatory genes by the colonic mucosa, but it is sufficient to expand colonic iNKT cells and imprint them toward an inflammatory profile.

We observed that upon antibiotic treatment, only colonic iNKT cells, but not those residing in the mLN or in the spleen, increased in frequency. Alterations in the survival of colonic iNKT cells or the incremented availability of colonic CXCL16 might be associated to iNKT cells frequency increase upon antibiotic treatment. In addition, this observation can be explained by the notion that iNKT cell proliferative capacity is negatively regulated by the presence in the intestinal microflora of the commensal microbe B. fragilis [22]. In GF mice, where B. fragilis is absent, mucosal iNKT cells expand without control [22]. Similarly, broad-spectrum antibiotics induce a rapid and significant drop in taxonomic richness and diversity, including a reduction in Bacteroidetes (28, 29), thus suggesting a similar explanation for the observed iNKT cell frequency increase in ABX-treated mice. Re-colonization of the gastrointestinal tract with a Bacteroidetes-rich healthy microbiota explains the observed iNKT cells frequency normalization upon eubiotic bacterial reconstitution.

Colonic conventional CD4+ T cells are also influenced by the presence or absence of specific bacterial strains [4, 5]. GF mice manifest impairments in CD4+ T cell functions (1, 30) while Vancomycin administration, selectively ablating Gram-negative bacteria, causes reduction of Tregs in colonic LP (5, 31). In our model, we did not observe variations in the frequency of CD4+ T helper (Th) cells, neither in the colon nor in the mesenteric LN or spleen, after 2 weeks broad-spectrum antibiotic administration. On the contrary, it was previously shown that a longer ABX treatment (8 weeks) could affect CD4+ and CD8+ T cells intestinal frequencies (16), suggesting that a shorter antibiotic exposure may not be sufficient to induce changes in the expansion of the MHC-restricted T cell population, as instead observed for non-classical lipid-specific iNKT cells.

We also observed that ABX treatment induced a skewing toward a pro-inflammatory cytokine profile in both colonic iNKT cells and conventional CD4+ T cells. No data were available so far on cytokine profile skewing upon short-term antibiotic treatment in colonic T cells. It was reported that a cocktail of ABX for 2 weeks was capable to skew the cytokine profile of systemic T cells promoting Th2 responses (32), similarly to short-term Kanamycin administration in 3-week-old mice [33]. A longer antibiotic treatment, instead, completely abrogated the overall cytokine production by intestinal CD4+ T cells (16), supporting the hypothesis that the duration of antibiotic treatment might differentially impact on the cytokine producing capability of intestinal T cells.

It has also been shown that re-association of antibiotic-treated mice with distinct bacterial species could alter the cytokine profile of conventional T cells (33). Enterococcus faecalis and Lactobacillus acidophilus could revert or attenuate the cytokine skewing of conventional Th subsets, while Bacteroides vulgatus caused exacerbation of Th-2 inflammatory responses [33].

Intestinal inflammation induces profound alterations of the gut microbiota ecosystem (34). IBD patients harbor an important bacterial diversity as compared to not-IBD controls, defined by an increase in Proteobacteria (such as E. coli adherent invasive and Enterobacteriaceae in CD) and a decrease in Firmicutes (such as F. prausnizii) [35]. Also, DSS-treated mice harbor a dysbiotic microbiota enriched in pathobionts [25], namely commensal microorganisms that bear the potential to cause pathology. These types of microorganisms are enriched in human IBD (36) and in murine models of intestinal inflammation [37] and when transferred into GF are sufficient to induce experimental intestinal inflammation [4, 37].

In the context of antibiotic-premedicated mice, we observed that re-colonization of the gastrointestinal tract with dysbiotic microbiota, but not with eubiotic microbiota, directly affected the activation status of iNKT cells (and of conventional CD4+ T cells), as demonstrated by their upregulation of CD69 and their skew toward a Th1/Th17 pro-inflammatory cytokine phenotype. At present, it remains to be elucidated whether iNKT cell activation depends on the recognition of stimulatory lipid antigens or rather on innate signals originated from bacterial pattern recognition molecules.

In this work, we observed that the functional imprinting of iNKT cells and of CD4+ T cells by a dysbiotic microbiota toward an activated/inflammatory Th1-Th17 cytokine profile after antibiotic treatment had important consequences when mice experienced subsequent intestinal inflammation. The transferred dysbiotic microbiota utilized to re-colonize mice showed a higher colitogenic potential than eubiotic microbiota in mice premedicated with antibiotics, and both iNKT cells and CD4+ T cells manifested a sustained Th1/Th17 skewed cytokine profile in this experimental condition. Indeed, in line with previous reports [38], this observation suggests that antibiotic treatment might alter the colonization niche, favoring the selective engraftment of pathogenic bacteria.

Induction of a pro-inflammatory phenotype by colonic T cells after antibiotic treatment by a dysbiotic microbiota might potentially have clinically relevant consequences. Patients that are genetically susceptible to harbor a dysbiotic microbiota, like those suffering from IBD, during the course of their disease might undergo treatments with antibiotics and experience episodes of intestinal inflammation. Indeed, current guidelines of ECCO [39, 40] do not recommend antibiotic treatment for IBD patients, unless infective complications are suspected or ongoing and before surgical interventions.

As observed in our experimental model, upon cessation of antibiotic treatment, IBD patients could re-colonize their gastrointestinal tract with a dysbiosis-prone microbiota, capable to induce Th1/Th17 secreting colonic (iNK)T cells and bearing an activated phenotype. Thus, the synergistic action of antibiotics and the dysbiotic microbiota might potentially aggravate the outcomes of IBD-related inflammation.

Our experimental data suggest that, at least in the absence of genetically predisposing factors, a longer time between microbiota engraftment and DSS-induced intestinal inflammation is sufficient to reduce the colitogenic potential of the dysbiotic microbiota. If this effect is secondary to the filling of the colonization niche by normal eubiotic bacteria remains to be elucidated. Nonetheless, it opens the possibility to evaluate therapeutic interventions, for example, by administering probiotics, to contrast the engraftment of pathobionts after antibiotic treatment.

Summary and Conclusion

In this study, we observed that antibiotic treatment profoundly altered the frequency of intestinal iNKT cells and the functions of both iNKT and CD4+ T cells even in the absence of concomitant intestinal inflammation. The presence of a dysbiotic microbiota after antibiotic treatment imprints colonic (iNK)T cells toward a pro-inflammatory phenotype that contributes to aggravate intestinal inflammation. Nonetheless, we observed that the inflammatory potential of the dysbiotic microbiota decreased over time, at least in a system uncoupled with genetic predisposition to select and maintain the engraftment of pathobionts. At present, it remains to be elucidated whether (iNK)T cells activation in these conditions depends on the recognition of stimulatory lipidic or proteic antigens or rather on innate signals conveyed by the dysbiotic microbiota.

In immune-mediated intestinal pathologies, it has not been fully elucidated whether intestinal dysbiosis may be the cause or a consequence of intestinal inflammation. However, transfer of dysbiotic microflora or single commensal bacterial species in GF is sufficient to induce experimental intestinal inflammation and activate the mucosal immune system [4, 37]. What it is now emerging is that antibiotic treatment can also induce dysbiosis and, as we observed, this can promote activating pro-inflammatory immune cell responses in the colon.

Individuals with genetic predispositions to harbor a dysbiotic microbiota and to aberrantly activate the mucosal immune system are, therefore, more exposed to unwanted pro-inflammatory immune responses after antibiotic treatment.

Albeit antibiotic treatment cannot be avoided, it should be kept in mind to predispose, especially for those individuals, treatments to re-equilibrate antibiotic-induced dysbiosis. For example, probiotics administration could contribute to contrast pathobionts engraftment and simultaneously switching-off mucosal immune cell activations.

References

1. Macpherson AJ, Harris NL. Interactions between commensal intestinal bacteria and the immune system. Nat Rev Immunol (2004) 4(6):478–85. doi:10.1038/nri1373

2. Sommer F, Backhed F. The gut microbiota – masters of host development and physiology. Nat Rev Microbiol (2013) 11(4):227–38. doi:10.1038/nrmicro2974

3. Becattini S, Taur Y, Pamer EG. Antibiotic-induced changes in the intestinal microbiota and disease. Trends Mol Med(2016) 22(6):458–78. doi:10.1016/j.molmed.2016.04.003

4. Ivanov II, Atarashi K, Manel N, Brodie EL, Shima T, Karaoz U, et al. Induction of intestinal Th17 cells by segmented filamentous bacteria. Cell (2009) 139(3):485–98. doi:10.1016/j.cell.2009.09.033

5. Atarashi K, Tanoue T, Oshima K, Suda W, Nagano Y, Nishikawa H, et al. Treg induction by a rationally selected mixture of Clostridia strains from the human microbiota. Nature (2013) 500(7461):232–6. doi:10.1038/nature12331

6. Petersen C, Round JL. Defining dysbiosis and its influence on host immunity and disease. Cell Microbiol (2014) 16(7):1024–33. doi:10.1111/cmi.12308

7. Levy M, Kolodziejczyk AA, Thaiss CA, Elinav E. Dysbiosis and the immune system. Nat Rev Immunol (2017) 17(4):219–32. doi:10.1038/nri.2017.7

8. Ianiro G, Tilg H, Gasbarrini A. Antibiotics as deep modulators of gut microbiota: between good and evil. Gut (2016) 65(11):1906–15. doi:10.1136/gutjnl-2016-312297

9. Buffie CG, Jarchum I, Equinda M, Lipuma L, Gobourne A, Viale A, et al. Profound alterations of intestinal microbiota following a single dose of clindamycin results in sustained susceptibility to Clostridium difficile-induced colitis. Infect Immun (2012) 80(1):62–73. doi:10.1128/IAI.05496-11

10. Hill DA, Siracusa MC, Abt MC, Kim BS, Kobuley D, Kubo M, et al. Commensal bacteria-derived signals regulate basophil hematopoiesis and allergic inflammation. Nat Med (2012) 18(4):538–46. doi:10.1038/nm.2657

11. Russell SL, Gold MJ, Hartmann M, Willing BP, Thorson L, Wlodarska M, et al. Early life antibiotic-driven changes in microbiota enhance susceptibility to allergic asthma. EMBO Rep (2012) 13(5):440–7. doi:10.1038/embor.2012.32

12. Ungaro R, Bernstein CN, Gearry R, Hviid A, Kolho KL, Kronman MP, et al. Antibiotics associated with increased risk of new-onset Crohn's disease but not ulcerative colitis: a meta-analysis. Am J Gastroenterol (2014) 109(11):1728–38. doi:10.1038/ajg.2014.246

13. Ananthakrishnan AN, Bernstein CN, Iliopoulos D, Macpherson A, Neurath MF, Ali RAR, et al. Environmental triggers in IBD: a review of progress and evidence. Nat Rev Gastroenterol Hepatol (2017) 15(1):39–49. doi:10.1038/nrgastro.2017.136

14. Brandl K, Plitas G, Mihu CN, Ubeda C, Jia T, Fleisher M, et al. Vancomycin-resistant enterococci exploit antibiotic-induced innate immune deficits. Nature (2008) 455(7214):804–7. doi:10.1038/nature07250

15. Atarashi K, Tanoue T, Shima T, Imaoka A, Kuwahara T, Momose Y, et al. Induction of colonic regulatory T cells by indigenous Clostridium species. Science (2011) 331(6015):337–41. doi:10.1126/science.1198469

16. Ekmekciu I, von Klitzing E, Fiebiger U, Escher U, Neumann C, Bacher P, et al. Immune responses to broad-spectrum antibiotic treatment and fecal microbiota transplantation in mice. Front Immunol (2017) 8:397. doi:10.3389/fimmu.2017.00397

M, Porcelli SA. CD1-restricted T cells in host defense to infectious diseases. Curr Top Microbiol Immunol (2007) 314:215–50.

18. Tupin E, Kinjo Y, Kronenberg M. The unique role of natural killer T cells in the response to microorganisms. Nat Rev Microbiol (2007) 5(6):405–17. doi:10.1038/nrmicro1657

19. Facciotti F, Ramanjaneyulu GS, Lepore M, Sansano S, Cavallari M, Kistowska M, et al. Peroxisome-derived lipids are self antigens that stimulate invariant natural killer T cells in the thymus. Nat Immunol (2012) 13(5):474–80. doi:10.1038/ni.2245

20. Bendelac A, Savage PB, Teyton L. The biology of NKT cells. Annu Rev Immunol (2007) 25:297–336. doi:10.1146/annurev.immunol.25.022106.141711

21. Nieuwenhuis EE, Matsumoto T, Lindenbergh D, Willemsen R, Kaser A, Simons-Oosterhuis Y, et al. Cd1d-dependent regulation of bacterial colonization in the intestine of mice. J Clin Invest (2009) 119(5):1241–50. doi:10.1172/JCI36509

22. Olszak T, An D, Zeissig S, Vera MP, Richter J, Franke A, et al. Microbial exposure during early life has persistent effects on natural killer T cell function. Science (2012) 336(6080):489–93. doi:10.1126/science.1219328

23. An D, Oh SF, Olszak T, Neves JF, Avci FY, Erturk-Hasdemir D, et al. Sphingolipids from a symbiotic microbe regulate homeostasis of host intestinal natural killer T cells. Cell (2014) 156(1–2):123–33. doi:10.1016/j.cell.2013.11.042

24. Geissmann F, Cameron TO, Sidobre S, Manlongat N, Kronenberg M, Briskin MJ, et al. Intravascular immune surveillance by CXCR6+ NKT cells patrolling liver sinusoids. PLoS Biol (2005) 3(4):e113. doi:10.1371/journal.pbio.0030113

25. Munyaka PM, Rabbi MF, Khafipour E, Ghia JE. Acute dextran sulfate sodium (DSS)-induced colitis promotes gut microbial dysbiosis in mice. J Basic Microbiol (2016) 56(9):986–98. doi:10.1002/jobm.201500726

26. Paroni M, Maltese V, De Simone M, Ranzani V, Larghi P, Fenoglio C, et al. Recognition of viral and self-antigens by TH1 and TH1/TH17 central memory cells in patients with multiple sclerosis reveals distinct roles in immune surveillance and relapses. J Allergy Clin Immunol (2017) 140(3):797–808. doi:10.1016/j.jaci.2016.11.045

27. Human Microbiome Project Consortium. Structure, function and diversity of the healthy human microbiome. Nature(2012) 486(7402):207–14. doi:10.1038/nature11234

28. Dethlefsen L, Huse S, Sogin ML, Relman DA. The pervasive effects of an antibiotic on the human gut microbiota, as revealed by deep 16S rRNA sequencing. PLoS Biol (2008) 6(11):e280. doi:10.1371/journal.pbio.0060280

29. Dethlefsen L, Relman DA. Incomplete recovery and individualized responses of the human distal gut microbiota to repeated antibiotic perturbation. Proc Natl Acad Sci U S A (2011) 108(Suppl 1):4554–61. doi:10.1073/pnas.1000087107

30. Macpherson AJ, Hunziker L, McCoy K, Lamarre A. IgA responses in the intestinal mucosa against pathogenic and non-pathogenic microorganisms. Microbes Infect (2001) 3(12):1021–35. doi:10.1016/S1286-4579(01)01460-5

31. Tanoue T, Atarashi K, Honda K. Development and maintenance of intestinal regulatory T cells. Nat Rev Immunol (2016) 16(5):295–309. doi:10.1038/nri.2016.36

32. Dimmitt RA, Staley EM, Chuang G, Tanner SM, Soltau TD, Lorenz RG. Role of postnatal acquisition of the intestinal microbiome in the early development of immune function. J Pediatr Gastroenterol Nutr (2010) 51(3):262–73. doi:10.1097/MPG.0b013e3181e1a114

33. Sudo N, Yu XN, Aiba Y, Oyama N, Sonoda J, Koga Y, et al. An oral introduction of intestinal bacteria prevents the development of a long-term Th2-skewed immunological memory induced by neonatal antibiotic treatment in mice. Clin Exp Allergy (2002) 32(7):1112–6. doi:10.1046/j.1365-2222.2002.01430.x

34. Kaser A, Zeissig S, Blumberg RS. Inflammatory bowel disease. Annu Rev Immunol (2010) 28:573–621. doi:10.1146/annurev-immunol-030409-101225

35. Gevers D, Kugathasan S, Denson LA, Vazquez-Baeza Y, Van Treuren W, Ren B, et al. The treatment-naive microbiome in new-onset Crohn's disease. Cell Host Microbe (2014) 15(3):382–92. doi:10.1016/j.chom.2014.02.005

36. Frank DN, St Amand AL, Feldman RA, Boedeker EC, Harpaz N, Pace NR. Molecular-phylogenetic characterization of microbial community imbalances in human inflammatory bowel diseases. Proc Natl Acad Sci U S A (2007) 104(34):13780–5. doi:10.1073/pnas.0706625104

37. Garrett WS, Lord GM, Punit S, Lugo-Villarino G, Mazmanian SK, Ito S, et al. Communicable ulcerative colitis induced by T-bet deficiency in the innate immune system. Cell (2007) 131(1):33–45. doi:10.1016/j.cell.2007.08.017

38. Wlodarska M, Willing B, Keeney KM, Menendez A, Bergstrom KS, Gill N, et al. Antibiotic treatment alters the colonic mucus layer and predisposes the host to exacerbated Citrobacter rodentium-induced colitis. Infect Immun (2011) 79(4):1536–45. doi:10.1128/IAI.01104-10

39. Dignass A, Van Assche G, Lindsay JO, Lemann M, Soderholm J, Colombel JF, et al. The second European evidence-based consensus on the diagnosis and management of Crohn's disease: current management. J Crohns Colitis (2010) 4(1):28–62. doi:10.1016/j.crohns.2010.07.001

40. Dignass A, Lindsay JO, Sturm A, Windsor A, Colombel JF, Allez M, et al. Second European evidence-based consensus on the diagnosis and management of ulcerative colitis part 2: current management. J Crohns Colitis (2012) 6(10):991–1030. doi:10.1016/j.crohns.2012.09.002.

CHAPTER 3

Factors Promoting Development of Fibrosis in Crohn's Disease

The concepts on the pathophysiology of intestinal fibrosis in Crohn's disease (CD) have changed in recent years. Some years ago fibrosis was regarded to be a consequence of long-standing inflammation with subsequent destruction of the gut wall matrix followed by scar formation and collagen deposition. Fibrosis in CD patients appeared to be an irreversible process that could hardly be influenced. Therefore, the main target in CD therapy was to control inflammation to avoid fibrosis development. Many of these assumptions seem to be only partially true. Inflammation may be a necessary prerequisite for the initiation of fibrosis. However, when the pathophysiologic processes that lead to fibrosis in CD patients have been initiated fibrosis development may be independent of inflammation and may continue even when inflammation is under good medical control. Fibrosis in CD also may be reversible. After strictureplasty local collagen deposits decrease or even disappear. With new animal models for intestinal fibrosis on the horizon, we need to spend more efforts on understanding the factors influencing fibrosis in CD patients to finally find specific therapies. In this context, it will be as important to find markers and quantitative imaging tools to have reliable endpoints for clinical trials in fibrosing CD.

Introduction

Fibrosis in general can be characterized as exaggerated accumulation of collagen-rich extracellular matrix (ECM) in a tissue normally containing much less connective tissue with permanent or transient local expansion of mesenchymal cells or mesenchymal like cells and subsequent impairment of organ function [1].

Traditionally, fibrosis in Crohn's disease (CD) has been seen as a relatively slow process needing many months to develop [2]. In the discussion of delayed diagnosis of CD, it is usually emphasized that stricturing complications of CD and severe fibrosis of CD intestine could by avoided by a timely diagnosis [3]. However, recent data indicate that this may not necessarily be the case. A rapid development of fibrosis in some patients seems to be possible. Rapid reoccurrence of fibrosis has been described in patients that undergo liver transplantation for hepatic fibrosis or cirrhosis [4]. Rapid lung fibrosis could be induced by inhalative toxins in animal models [5, 6], and rapid liver fibrosis is seen in some models of primary sclerosing cholangitis [7]. These data and further evidence support the concept that under certain circumstances fibrosis and subsequent stricture formation in some CD patients may be much faster than traditionally assumed.

It is evident from clinical findings that fibrosis only develops in segments of the gut where inflammation in the context of CD is present [1]. Fibrosis in gut segments that never showed inflammatory involvement has not been reported. While this seems to be obvious, it is less clear what factors really trigger the process.

Another "dogma" also has been revised recently. It is no longer believed that only primary mesenchymal cells such as fibroblasts or smooth muscle cells can contribute to fibrosis in CD (1). Cells that contribute to fibrosis in CD patients may also derive from intestinal epithelial cells via a process called epithelial-to-mesenchymal transition (EMT) [1, 2, 8, 9] or from endothelial cells via endothelial-to-mesenchymal transition (EndoMT) [10].

A third important new aspect in the discussion is the assumption that fibrosis in CD may not be irreversible [11, 12]. After strictureplasty in patients with CD suffering from clinical strictures the fibrosis in the gut wall was later on found to be reduced or even completely absent (13). Reversibility of fibrosis had been demonstrated before in other fibrotic diseases such as liver fibrosis [14].

Intestinal fibrosis subsequently is neither necessarily a very slow process nor completely dependent on the presence of inflammation, nor irreversible [1, 2, 9, 12, 15]. Therefore, it appears to be important to review the cellular and molecular factors that contribute to fibrogenesis in CD.

Factors Activating Matrix-Producing Cells

Matrix-producing cells are activated by paracrine signals, autocrine factors, and pathogen-associated molecular patterns derived from microorganisms or damage-associated molecular patterns that interact with pattern recognition receptors [1, 2, 12, 15]. Transforming growth factor β (TGF-β) is an important mediator of mesenchymal cell activation. Its important role as a central regulator of fibrosis has been emphasized for many tissues and diseases [16–24]. TGF-β expression is found to be upregulated in inflamed mucosa of inflammatory bowel disease patients [25–28]. In addition, also inhibitory molecules of TGF-β action such as SMAD7 are upregulated in CD mucosa [29, 30]. Recent therapeutic approaches now target SMAD7 expression by an antisense oligonucleotide (Mongersen) to allow more TGF-β action mainly of regulatory T-cells [31]. It will be interesting to see whether a parallel activation of mesenchymal cells can be prevented [32]. Data on mesenchymal cell activation and collagen deposition derived from clinical trials that are under way with Mongersen will help us to understand which role TGF-β plays for the activation of mesenchymal cells, for the initiation of EMT or EndoMT and for gut wall fibrosis in CD patients.

Other factors that play an important role in activating mesenchymal cells are activins [33], connective tissue growth factor [34–36], platelet-derived growth factor, insulin-like growth factor (IGF-1, -2), epidermal growth factor, and endothelins (ET-1, -2, -3) [2, 12, 15]. All of those factors increase collagen synthesis by mesenchymal cells upon stimulation [2, 12, 15]. The relative contribution of the respective factors and whether synergies are developed is unclear. Therefore, it is also unclear whether targeting one of those factors in an anti-fibrotic therapeutic approach would make sense [1].

Besides those specific factors inflammation per se is a strong activator of mesenchymal cells and also contributes to EMT and EndoMD (1, 8). Therefore, it has been assumed by many authors that control of inflammation would prevent the development of gut wall fibrosis. This seems to be questionable now. Recent epidemiological data indicate that biologicals have reduced the number of surgeries performed due to insufficient control of inflammation. However, despite effective and much better control of inflammation the development of CD in general from a B1 phenotype (only inflammatory) to a B2 (fibrotic) or B3 (penetrating) phenotype seems not to be significantly reduced [37, 38]. This raises the important questions whether inflammatory mediators and molecules trigger the fibrotic process early and whether this process finally becomes independent from inflammation. If this would be the case—and there is quite some evidence to support this assumption—the development of anti-fibrotic therapies would be absolutely mandatory. If we cannot interfere with the progression of fibrosis in a significant number of patients with our current therapeutic armamentarium, the need for new drug development becomes obvious.

Animal Models to Study Fibrosis-Promoting Factors and Potential Therapies

Several animal models for the study of intestinal fibrosis have been proposed and described [1]. All of them have some advantages as well as disadvantages and none of them really resembles intestinal fibrosis of CD patients. Spontaneous intestinal fibrosis does not occur in rodent models, and therefore all models require some manipulation and artificial conditions.

The first models used to study intestinal fibrosis were models in which colonic inflammation was chemically induced, such as the trinitrobenzene-sulfonic acid (TNBS) and chronic dextran sodium sulfate (DSS) colitis in mice [1]. Some collagen deposition and fibrosis is observed in these models. However, fibrosis is usually inconsistent, and the experimental duration until the occurrence of fibrosis limits the applicability of the mentioned mouse models. In addition, the contribution of the chemical trigger of the inflammation (TNBS or DSS) raises some concerns with respect to pathophysiological relevance. Similar to those chemically triggered models, the injection of the bacterial wall-derived compound peptidoglycan–polysaccharide into the gut wall induces inflammation and fibrosis [39]. While this model is an example for fibrosis triggered by microbial products, it is unclear whether bacteria play an essential role in CD fibrosis. The SAMP1/Yit mouse was reported to develop spontaneous inflammation with ileitis and fibrosis [40, 41]. However, the access to this model is limited, and the extent of fibrosis seems to depend on the vivarium the mice are bred in.

To be able to study therapeutic interventions with the target of inhibiting intestinal fibrosis we established a new—but still very artificial model. For the study of bronchiolitis obliterans and bronchial fibrosis, pulmonologists had developed a heterotopic transplant model of trachea in rats [42]. We adopted this model and investigated whether the heterotopic transplantation of small intestine into the neck fold of rats would also be followed by the development of fibrosis. Indeed, we detected a rapid fibrosis of the small intestinal wall occurring within 2 weeks [43]. This was associated with increased expression of typical mediators of fibrosis such as αvβ6 integrin, IL-13, and TGF-β [43]. Further, we detected a loss of intestinal epithelium morphology, could demonstrate exaggerated collagen deposition, which led to luminal wall thickening culminating in a veritable fibrotic occlusion of the intestinal lumen. As the available reagents to study fibrosis in rats are limited and it was desirably to study certain knockout or transgenic animal models we investigated whether the heterotopic transplant model also would work in mice (Figures 1A–C). As expected, we found a similar time course of development of fibrosis in the mouse model (44). C57BL/6 mice are used as donors for isogeneic transplantation into C57BL/6 recipients in this model. Interestingly, a rapid revascularization occurs in the intestinal grafts in the neck fold (Figures 1D–F). In small intestinal grafts isolated up to 21 days after transplantation, the lumen was obstructed by granulation tissue and fibrotic material (Figures 1G–J [44]). The grafts partially had lost their typical crypt structure which in some specimen occurred already at day 2 after transplantation indicating that hypoxia may have an important role for this development.

Collagen layer thickness was observed to be significantly increased in grafts in a time-dependent manner [Figures 2A,B; 44]. Confirmatively, Tgf-β and collagen mRNA was observed to be significantly increased in a time-dependent manner [Figures 2C–E; 44].

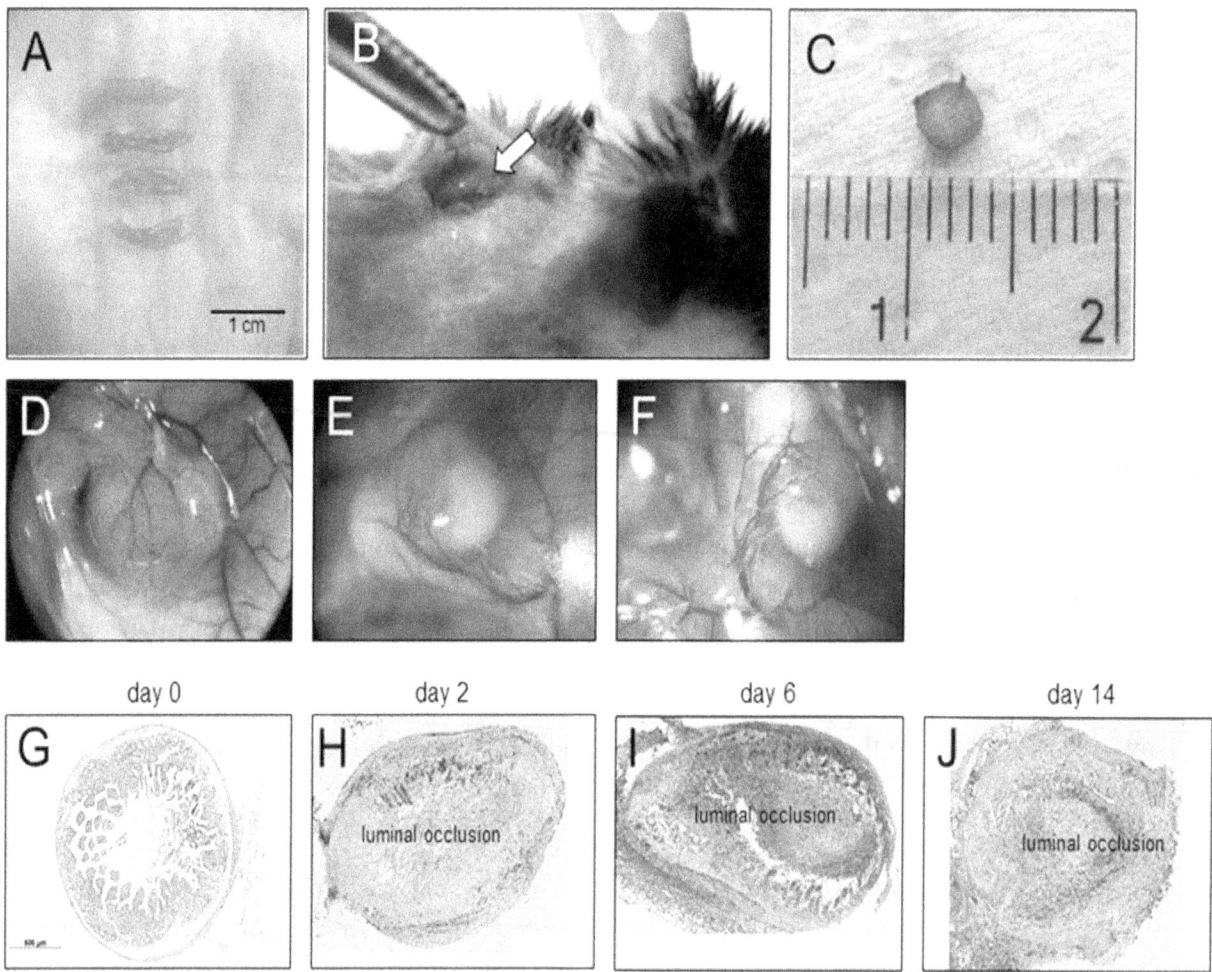

Figure 1. Heterotopic transplantation, revascularization, and luminal occlusion of the graft. (A) Small bowel resections are extracted from C57BL/6 mice. (B) For isogeneic transplantation, the resection (arrow) is implanted into subcutaneous tissue in the neck of C57BL/6 mice. (C) The graft is freed from the pouch and harvested from the neck of the recipient 14 days posttransplantation. (D–F) Grafts in the neck of recipient animals observed in situ present a decreased length but are otherwise macroscopically intact. Blood vessels from the surrounding tissue stretch toward the graft where they form a dense network (twofold magnification). (G) Histologic cross sections of freshly isolated small intestine (day 0). Small bowel resections are extracted from C57BL/6 mice, implanted into C57BL/6 mice for isogeneic transplantation, and explanted at (H) day 2, (I) day 6, and (J) day 14 after transplantation. Transmitted light microscopy, H&E staining. Grafts revealed luminal occlusion.

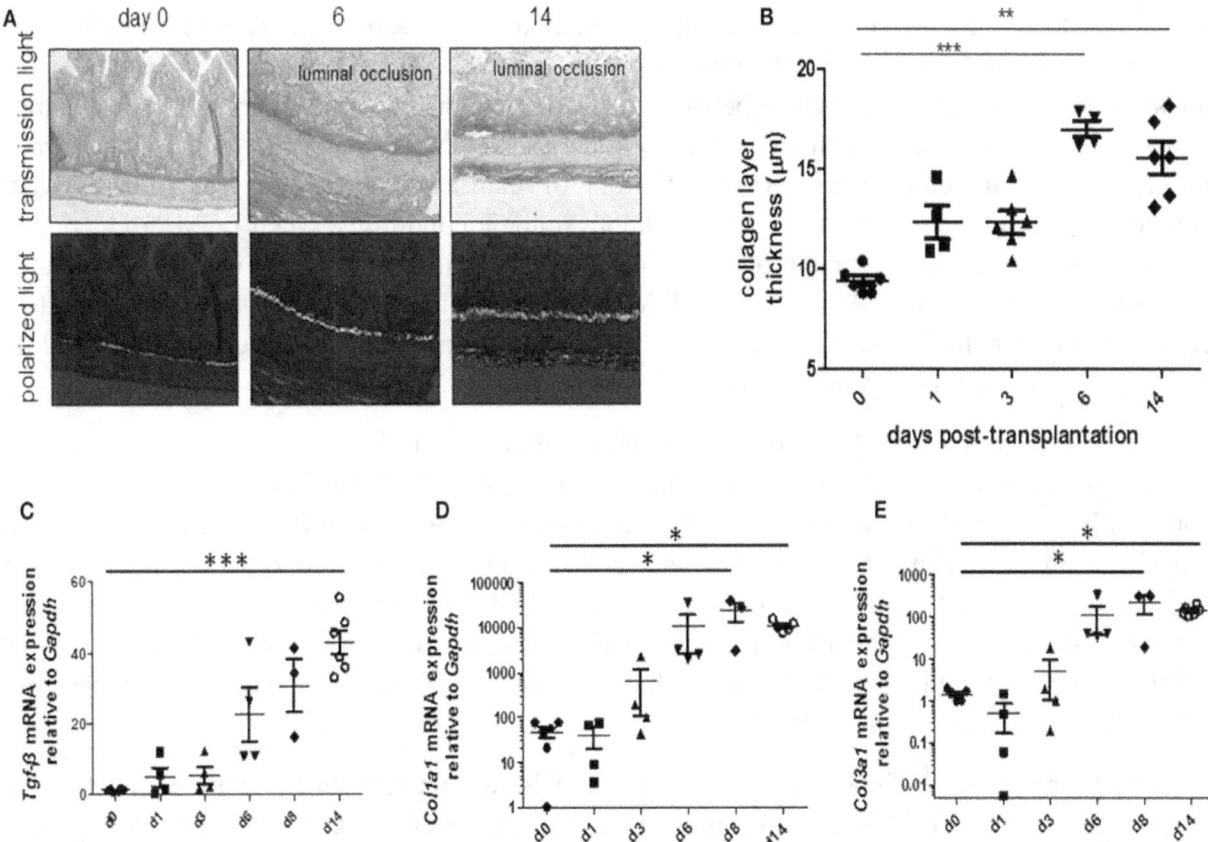

Figure 2. Collagen layer thickness and Tgf-β, Col1a1, and Col3a1 mRNA are significantly increased in grafts from the heterotopic transplantation model in a time-dependent manner. Small bowel resections are extracted from RAG2 knockout mice and implanted into RAG2 knockout mice for isogeneic heterotopic transplantations. (A) Sirius Red staining. Transmission light microscopy and polarized light microscopy. (B) Collagen layer thickness measurement using transmission light microscopy confirmed significantly increased collagen layer thickness in a time-dependent manner (**p < 0.01, ***p < 0.001, ANOVA, Dunn's multiple comparison test). Thickness was calculated from at least eight places in representative areas at 10-fold magnification for each single graft. (C) Tgf-β qPCR. (D) Col1a1. (E) Col1a3. *p < 0.05, ***p < 0.001, Kruskal–Wallis test, Dunn's multiple comparison test.

We used this newly developed model to study established anti-fibrotic drugs and their effect on the development of fibrosis [44]. Pirfenidone so far is the best established therapy for idiopathic lung fibrosis [45–47]. When we applied pirfenidone three times a day for 6 days by oral gavage, we found that the collagen layer was significantly decreased in comparison to the collagen layer thickness in grafts from vehicle treated mice (44). Similar, TGF-β mRNA expression was significantly decreased upon pirfenidone treatment compared to vehicle [44].

Factors Involved in Tissue Remodeling

Additional factors involved in intestinal fibrosis that have not been discussed so far regulate the turnover of the ECM [2, 12, 15, 48]. It is generally assumed that in normal tissue, i.e., in the normal intestinal wall there is a fine balance between ECM production and degradation (1). This balance is maintained on one hand by matrix metalloproteinases (MMPs) that break down and degrade ECM, and on the other hand tissue inhibitors of matrix metalloproteinases (TIMPs) that counteract this degrading activity. Under pathophysiologic conditions, when ECM production is increased and surpasses degradation intestinal fibrosis will occur. In human, CD strictures increased expression of MMPs, and also TIMPs has been observed. However, it is difficult of course to functionally investigate the balance and dynamics between the different pro-degrading and degradation-inhibiting proteins and mechanisms.

Further functions of MMP-9 include the regulation of cell migration, invasion, cell signaling as well as induction and regulation of EMT in multiple tissues [49–51]. In fact, MMP-9 is the most abundantly expressed tissue degrading and remodeling protease in inflamed CD tissue [52]. In biopsies from CD patients, MMP-9 was found as latent (pro-) and mature form [53]. Further, serum and urinary levels of MMP-9 correlate with disease activity in CD patients. It has been suggested that MMP-9 serum levels could be a useful marker of CD disease activity in children. In DSS colitis in mice, targeted deletion of MMP-9 has a protective effect, whereas mice overexpressing MMP-9 develop more severe colitis [53].

As MMPs are obviously involved in intestinal fibrosis, we determined the expression of tissue remodeling proteases MMP-2, -9, -13, and TIMP-1 in our heterotopic transplant model by real-time PCR. When mice were treated with pirfenidone, a significant decrease in MMP-9 mRNA expression was observed [44]. Similar, MMP-2, -13, and TIMP-1 mRNA expression was decreased upon pirfenidone [44].

Further, we investigated, whether the therapeutic neutralization of MMP-9 by specific antibodies would alter the development of fibrosis in the heterotopic transplant model. When we treated mice in our model with two different anti-MMP-9 antibodies, the lumen of the intestinal grafts was only partially obstructed, and some crypt structures were still present [53]. Whereas the collagen layer was much thicker in grafts harvested from the isotype control-treated group, grafts harvested from anti-MMP-9 antibody-treated mice showed almost "normal" collagen layer thickness [53]. Treatment with the two anti-MMP-9 antibodies was followed by lower accumulation of newly synthesized collagen, significantly thinner collagen layer, and lower collagen-specific amino acid hydroxyproline. Expression of MMP-9 and TIMP-1 were not significantly changed by the MMP-9 antibody treatment [53]. When we assessed gelatinase activity in homogenates from our grafts by zymography and by ELISA, all day-14 explants exhibited increased total MMP-9, and the MMP-9 antibody treatment was followed by some reduction of MMP-9 activity in the explants [53].

Summary

We have just started to more specifically understand the factors and pathways that lead to intestinal fibrosis. This is necessary to address the high clinical need of focused treatment of fibrosis in CD patients. New animal models may be helpful to screen for successful therapies. In some models, such as the heterotopic transplant model of small intestinal segments, pirfenidone and anti-MMP-9 antibodies have provided promising results. Further studies will be necessary to confirm these results and to find additional factors promoting development of fibrosis in CD.

References

1. Rieder F, Fiocchi C, Rogler G. Mechanisms, management, and treatment of fibrosis in patients with inflammatory bowel diseases. Gastroenterology (2017) 152:340–50.e6.

2. Latella G, Di Gregorio J, Flati V, Rieder F, Lawrance IC. Mechanisms of initiation and progression of intestinal fibrosis in IBD. Scand J Gastroenterol (2015) 50:53–65.

3. Schoepfer AM, Dehlavi MA, Fournier N, Safroneeva E, Straumann A, Pittet V, et al. Diagnostic delay in Crohn's disease is associated with a complicated disease course and increased operation rate. Am J Gastroenterol (2013) 108:1744–53; quiz 1754.

4. Vasavada BB, Chan CL. Rapid fibrosis and significant histologic recurrence of hepatitis C after liver transplant is associated with higher tumor recurrence rates in hepatocellular carcinomas associated with hepatitis C virus-related liver disease: a single center retrospective analysis. Exp Clin Transplant (2015) 13:46–50.

5. Dong J, Ma Q. TIMP1 promotes multi-walled carbon nanotube-induced lung fibrosis by stimulating fibroblast activation and proliferation. Nanotoxicology (2017) 11:41–51.

6. Dong J, Ma Q. Myofibroblasts and lung fibrosis induced by carbon nanotube exposure. Part Fibre Toxicol (2016) 13:60.

7. Ikenaga N, Liu SB, Sverdlov DY, Yoshida S, Nasser I, Ke Q, et al. A new Mdr2(-/-) mouse model of sclerosing cholangitis with rapid fibrosis progression, early-onset portal hypertension, and liver cancer. Am J Pathol (2015) 185:325–34.

8. Rieder F, Brenmoehl J, Leeb S, Scholmerich J, Rogler G. Wound healing and fibrosis in intestinal disease. Gut (2007) 56:130–9.

9. Rieder F, Fiocchi C. Intestinal fibrosis in IBD – a dynamic, multifactorial process. Nat Rev Gastroenterol Hepatol (2009) 6:228–35.

10. Rieder F, Kessler SP, West GA, Bhilocha S, de la Motte C, Sadler TM, et al. Inflammation-induced endothelial-to-mesenchymal transition: a novel mechanism of intestinal fibrosis. Am J Pathol (2011) 179:2660–73.

11. Latella G, Sferra R, Speca S, Vetuschi A, Gaudio E. Can we prevent, reduce or reverse intestinal fibrosis in IBD? Eur Rev Med Pharmacol Sci (2013) 17:1283–304.

12. Lawrance IC, Rogler G, Bamias G, Breynaert C, Florholmen J, Pellino G, et al. Cellular and molecular mediators of intestinal fibrosis. J Crohns Colitis (2015):1–13.

13. Yamamoto T, Fazio VW, Tekkis PP. Safety and efficacy of strictureplasty for Crohn's disease: a systematic review and meta-analysis. Dis Colon Rectum (2007) 50:1968–86.

14. Arthur MJ. Reversibility of liver fibrosis and cirrhosis following treatment for hepatitis C. Gastroenterology (2002) 122:1525–8.

15. Latella G, Rogler G, Bamias G, Breynaert C, Florholmen J, Pellino G, et al. Results of the 4th scientific workshop of the ECCO (I): pathophysiology of intestinal fibrosis in IBD. J Crohns Colitis (2014) 8(10):1147–65.

16. Loboda A, Sobczak M, Jozkowicz A, Dulak J. TGF-beta1/Smads and miR-21 in renal fibrosis and inflammation. Mediators Inflamm (2016) 2016:8319283.

17. Meng XM, Nikolic-Paterson DJ, Lan HY. TGF-beta: the master regulator of fibrosis. Nat Rev Nephrol (2016) 12:325–38.

18. Sutariya B, Jhonsa D, Saraf MN. TGF-beta: the connecting link between nephropathy and fibrosis. Immunopharmacol Immunotoxicol (2016) 38:39–49.

19. Xu F, Liu C, Zhou D, Zhang L. TGF-beta/SMAD pathway and its regulation in hepatic fibrosis. J Histochem Cytochem (2016) 64:157–67.

20. Munoz-Felix JM, Gonzalez-Nunez M, Martinez-Salgado C, Lopez-Novoa JM. TGF-beta/BMP proteins as therapeutic targets in renal fibrosis. Where have we arrived after 25 years of trials and tribulations? Pharmacol Ther (2015) 156:44–58.

21. Meng XM, Tang PM, Li J, Lan HY. TGF-beta/Smad signaling in renal fibrosis. Front Physiol (2015) 6:82.

22. Weiskirchen R, Meurer SK. BMP-7 counteracting TGF-beta1 activities in organ fibrosis. Front Biosci (Landmark Ed) (2013) 18:1407–34.

23. Samarakoon R, Overstreet JM, Higgins PJ. TGF-beta signaling in tissue fibrosis: redox controls, target genes and therapeutic opportunities. Cell Signal (2013) 25:264–8.

24. Biernacka A, Dobaczewski M, Frangogiannis NG. TGF-beta signaling in fibrosis. Growth Factors (2011) 29:196–202.

25. Li C, Flynn RS, Grider JR, Murthy KS, Kellum JM, Akbari H, et al. Increased activation of latent TGF-beta1 by alphaVbeta3 in human Crohn's disease and fibrosis in TNBS colitis can be prevented by cilengitide. Inflamm Bowel Dis (2013) 19:2829–39.

26. Li C, Iness A, Yoon J, Grider JR, Murthy KS, Kellum JM, et al. Noncanonical STAT3 activation regulates excess TGF-beta1 and collagen I expression in muscle of stricturing Crohn's disease. J Immunol (2015) 194:3422–31.

27. Scarpa M, Bortolami M, Morgan SL, Kotsafti A, Ruffolo C, D'Inca R, et al. TGF-beta1 and IGF-1 and anastomotic recurrence of Crohn's disease after ileo-colonic resection. J Gastrointest Surg (2008) 12:1981–90.

28. Del Zotto B, Mumolo G, Pronio AM, Montesani C, Tersigni R, Boirivant M. TGF-beta1 production in inflammatory bowel disease: differing production patterns in Crohn's disease and ulcerative colitis. Clin Exp Immunol (2003) 134:120–6.

29. Zorzi F, Calabrese E, Monteleone I, Fantini M, Onali S, Biancone L, et al. A phase 1 open-label trial shows that smad7 antisense oligonucleotide (GED0301) does not increase the risk of small bowel strictures in Crohn's disease. Aliment Pharmacol Ther (2012) 36:850–7.

30. Monteleone G, Fantini MC, Onali S, Zorzi F, Sancesario G, Bernardini S, et al. Phase I clinical trial of Smad7 knockdown using antisense oligonucleotide in patients with active Crohn's disease. Mol Ther (2012) 20:870–6.

31. Monteleone G, Neurath MF, Ardizzone S, Di Sabatino A, Fantini MC, Castiglione F, et al. Mongersen, an oral SMAD7 antisense oligonucleotide, and Crohn's disease. N Engl J Med (2015) 372:1104–13.

32. Danese S, Fiorino G, Peyrin-Biroulet L. Targeting SMAD7 in Crohn's disease by Mongersen: therapeutic revolution under way? Gastroenterology (2015) 149:1121–3.

33. Dignass AU, Jung S, Harder-d'Heureuse J, Wiedenmann B. Functional relevance of activin A in the intestinal epithelium. Scand J Gastroenterol (2002) 37:936–43.

34. Beddy D, Mulsow J, Watson RW, Fitzpatrick JM, O'Connell PR. Expression and regulation of connective tissue growth factor by transforming growth factor beta and tumour necrosis factor alpha in fibroblasts isolated from strictures in patients with Crohn's disease. Br J Surg (2006) 93:1290–6.

35. di Mola FF, Di Sebastiano P, Gardini A, Innocenti P, Zimmermann A, Buchler MW, et al. Differential expression of connective tissue growth factor in inflammatory bowel disease. Digestion (2004) 69:245–53.

36. Dammeier J, Brauchle M, Falk W, Grotendorst GR, Werner S. Connective tissue growth factor: a novel regulator of mucosal repair and fibrosis in inflammatory bowel disease? Int J Biochem Cell Biol (1998) 30:909–22.

37. Jeuring SF, van den Heuvel TR, Liu LY, Zeegers MP, Hameeteman WH, Romberg-Camps MJ, et al. Improvements in the long-term outcome of Crohn's disease over the past two decades and the relation to changes in medical management: results from the population-based IBDSL cohort. Am J Gastroenterol (2017) 112:325–36.

38. van den Heuvel TR, Jonkers DM, Jeuring SF, Romberg-Camps MJ, Oostenbrug LE, Zeegers MP, et al. Cohort profile: the inflammatory bowel disease South Limburg cohort (IBDSL). Int J Epidemiol (2015) 46(2):e7(1–9).

39. Sartor RB, Anderle SK, Rifai N, Goo DA, Cromartie WJ, Schwab JH. Protracted anemia associated with chronic, relapsing systemic inflammation induced by arthropathic peptidoglycan-polysaccharide polymers in rats. Infect Immun (1989) 57:1177–85.

40. Strober W, Nakamura K, Kitani A. The SAMP1/Yit mouse: another step closer to modeling human inflammatory bowel disease. J Clin Invest (2001) 107:667–70.

41. Pizarro TT, Pastorelli L, Bamias G, Garg RR, Reuter BK, Mercado JR, et al. SAMP1/YitFc mouse strain: a spontaneous model of Crohn's disease-like ileitis. Inflamm Bowel Dis (2011) 17:2566–84.

42. Gammie JS, Li S, Kawaharada N, Colson YL, Yousem S, Ildstad ST, et al. Mixed allogeneic chimerism prevents obstructive airway disease in a rat heterotopic tracheal transplant model. J Heart Lung Transplant (1998) 17:801–8.

43. Hausmann M, Rechsteiner T, Caj M, Benden C, Fried M, Boehler A, et al. A new heterotopic transplant animal model of intestinal fibrosis. Inflamm Bowel Dis (2013) 19:2302–14.

44. Meier R, Lutz C, Cosin-Roger J, Fagagnini S, Bollmann G, Hunerwadel A, et al. Decreased fibrogenesis after treatment with pirfenidone in a newly developed mouse model of intestinal fibrosis. Inflamm Bowel Dis (2016) 22:569–82.

45. Xaubet A, Serrano-Mollar A, Ancochea J. Pirfenidone for the treatment of idiopathic pulmonary fibrosis. Expert Opin Pharmacother (2014) 15:275–81.

46. Takeda Y, Tsujino K, Kijima T, Kumanogoh A. Efficacy and safety of pirfenidone for idiopathic pulmonary fibrosis. Patient Prefer Adherence (2014) 8:361–70.

47. King TE Jr, Bradford WZ, Castro-Bernardini S, Fagan EA, Glaspole I, Glassberg MK, et al. A phase 3 trial of pirfenidone in patients with idiopathic pulmonary fibrosis. N Engl J Med (2014) 370:2083–92.

48. Rieder F, Karrasch T, Ben-Horin S, Schirbel A, Ehehalt R, Wehkamp J, et al. Results of the 2nd scientific workshop of the ECCO (III): basic mechanisms of intestinal healing. J Crohns Colitis (2012) 6:373–85.

49. Kurzepa J, Madro A, Czechowska G, Kurzepa J, Celinski K, Kazmierak W, et al. Role of MMP-2 and MMP-9 and their natural inhibitors in liver fibrosis, chronic pancreatitis and non-specific inflammatory bowel diseases. Hepatobiliary Pancreat Dis Int (2014) 13:570–9.

50. Farina AR, Mackay AR. Gelatinase B/MMP-9 in tumour pathogenesis and progression. Cancers (Basel) (2014) 6:240–96.

51. Vandooren J, Van den Steen PE, Opdenakker G. Biochemistry and molecular biology of gelatinase B or matrix metalloproteinase-9 (MMP-9): the next decade. Crit Rev Biochem Mol Biol (2013) 48:222–72.

52. Kofla-Dlubacz A, Matusiewicz M, Krzystek-Korpacka M, Iwanczak B. Correlation of MMP-3 and MMP-9 with Crohn's disease activity in children. Dig Dis Sci (2012) 57:706–12.

53. Goffin L, Fagagnini S, Vicari A, Mamie C, Melhem H, Weder B, et al. Anti-MMP-9 antibody: a promising therapeutic strategy for treatment of inflammatory bowel disease complications with fibrosis. Inflamm Bowel Dis (2016) 22:2041–57.

www.ingramcontent.com/pod-product-compliance
Lightning Source LLC
Chambersburg PA
CBHW081228170526
45165CB00009B/2998